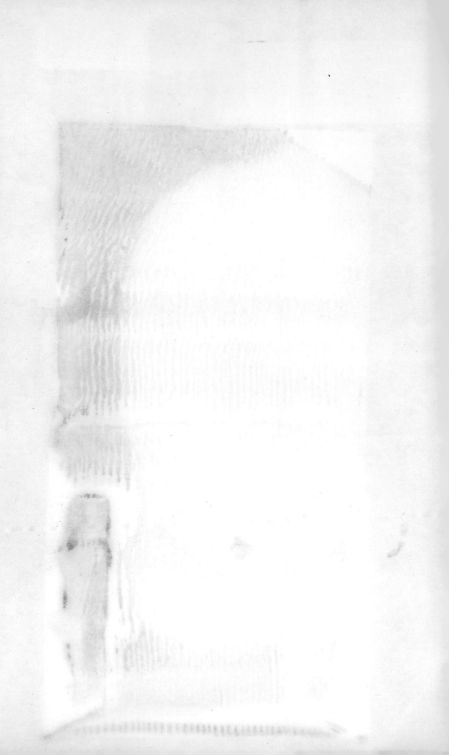

HISTORY, PHILOSOPHY AND SOCIOLOGY OF SCIENCE

SOCIOLOGY OF SCIENCE

Classics, Staples and Precursors

HISTORY, PHILOSOPHY AND SOCIOLOGY OF SCIENCE

Classics, Staples and Precursors

Selected By

YEHUDA ELKANA
ROBERT K. MERTON
ARNOLD THACKRAY
HARRIET ZUCKERMAN

THE DIFFUSION OF SCIENCE

By

JESSE LEE BENNETT

ARNO PRESS

A New York Times Company

New York — 1975

Reprint Edition 1975 by Arno Press Inc.

Reprinted from a copy in
    The St. Louis Public Library

HISTORY, PHILOSOPHY AND SOCIOLOGY OF SCIENCE:
Classics, Staples and Precursors
ISBN for complete set: 0-405-06575-2
See last pages of this volume for titles.

Manufactured in the United States of America

―――♦―――

Library of Congress Cataloging in Publication Data

Bennett, Jesse Lee, 1885-1931.
    The diffusion of science.

    (History, philosophy and sociology of science)
    Reprint of the 1942 ed. published by Johns Hopkins
Press, Baltimore.
    1. Communication in science. 2. Science--Study and
teaching. I. Title. II. Series.
Q223.B46 1975   507   74-26250
ISBN 0-405-06580-9

THE DIFFUSION OF SCIENCE

THE DIFFUSION
OF SCIENCE

By

JESSE LEE BENNETT

*Editor of " The Modern World," Author of " The Essential American
Tradition," " Frontiers of Knowledge," " What Books Can Do
For You," " On ' Culture' and A Liberal
Education, " etc.*

> "Since less than one out of each hundred of the
> findings of modern science are known to less
> than one out of each hundred persons of the
> general population, modern science has a social
> utility of less than one one-hundredth of one
> per cent."   —LESTER F. WARD.

BALTIMORE

THE JOHNS HOPKINS PRESS

1942

PRINTED IN THE UNITED STATES OF AMERICA
BY J. H. FURST COMPANY, BALTIMORE, MARYLAND

PREFACE

The accumulated results of the progress of science have
given the present generation such understanding of the
nature of life, of man, of society, and of the general environment of life, man, and society, as mankind has never
before possessed. The devices resulting from certain advances in pure science have given mankind unprecedented
control of environment and have permitted such complete
and immediate intercommunication among men as was
never before imagined. By some of these devices the distribution of physical things has been brought to a high
state of perfection in all civilized nations and, to less extent, internationally. But the devices of intercommunication have not yet effectively and successfully been used to
help diffuse among mankind the science of which they are,
themselves, the relatively incidental products.

The educational systems, inherited from a past in which
they evolved amid conditions very unlike those now existing to serve ends not identical with the ends which they
are now needed to serve have not been radically altered or
reshaped to deal with the new requirements and new potentialities of a new age. Scientific method has not yet been
used to study the question of the most efficient method of
diffusing the basic essentials of the new scientific knowledge among mankind by means of planned new educational instrumentalities and by use of the new devices of
intercommunication. While the vanguard of the race is in
possession of new, relatively exact, verifiable, demon-

strable, and communicable knowledge changing all previous conceptions of life, man, society, and " the universe," the great body of the race is still left at the mercy of ignorance, illusion, superstition, totem and tabu. Spasmodic and uncoordinated efforts to diffuse science prove relatively ineffectual largely because they are not made with articulate purpose and as parts of a carefully planned and continuous program. It grows ever more obvious that there is great unhappiness, much perplexity and many international and domestic, social, political, and economic aberrations which could be greatly diminished by the effective diffusion of the new scientific knowledge which we possess.

The application, on the physical level, of modern scientific knowledge results in improved health, increased span of life, and the rapid emergence from assorted physical miseries which have long harassed successive generations of mankind. There is every reason to believe that the application, on the mental level, of modern scientific knowledge would result in improved mental well-being, increased happiness, and the rapid emergence from assorted mental miseries which have eternally harassed mankind in its attempts to adapt itself happily and harmoniously to an intricate and perplexing environment and in its attempts to solve many questions to which, for the first time, we are now gaining satisfactory and irrefutable answers. The results gained in the Panama Canal Zone indicate what can be done by the application, on the physical level, of modern, scientific knowledge by centralized authority. No analogous centralized authority can hope to operate in the field of the diffusion of knowledge; but there are a number of great national and international bodies charged, by their charters, with " the advancement and diffusion of knowledge," which can at least clarify the new situation, formulate general purposes and plans and afford all those concerned in any way with agencies or instrumentalities through which information or ideas are disseminated, new understanding of and new perspective upon the changed conditions now confronting the world. Such activities might well serve to coordinate the present uncoordinated activities of endless small and great institutions. The cumulative results of such activities over a generation might be comparable with such results as were gained in the realm of physical hygiene in the Panama Canal Zone. Science can surely be diffused much more effectively. But such effective diffusion of science requires a plan and program based upon broad study of the changed world situation, of the nature of the new scientific knowledge, of the nature of the existing educational philosophy and educational systems, of the new devices of intercommunication, and of possible methods of deliberate and conscious procedure in a field in which existing activities are largely empirical and pragmatic.

This book will seek to envisage the present situation and the general problem very broadly and to offer one possible plan for the more effective diffusion of science under the new conditions which have so rapidly evolved. Very especially will this book plead for the application of scientific method to the whole problem of the diffusion of science. It was the application of scientific method to the problem of the advancement of science which disrupted the folkways of mediaevalism and brought the modern world into being. The application of scientific method to the problem of the diffusion of science might make that modern world exist in the minds of a larger proportion of mankind instead of merely in the form of startling and rather dangerous new devices manipulated by and influencing anachronistic mentalities.

JESSE LEE BENNETT

1930

CONTENTS

PAGE

I. The Great and Sudden World Changes.... 1

II. The Knowledge of the Past and the Agencies for Diffusing it 13

III. The New Scientific Knowledge and the Evolving Agencies for Diffusing it........... 39

IV. The Growing Realization of the Changed Situation 55

V. The Need for Broad View, Scientific Method, and Definite Plan.................... 68

VI. The Possible Aims of the Education of the Future 87

VII. The Possible Material of the Education of the Future 100

VIII. The Possible Technique of the Education of the Future 119

IX. Immediate Possibilities 128

Notes................................ 135

ix

I

THE GREAT AND SUDDEN WORLD CHANGES

Most of the conditions affecting the daily lives and thoughts of men have changed more completely during the past two decades than in any previous century. Between the conditions of life of a Maryland farmer, for example, in the year 1900 and the conditions of life of all rural dwellers down the ages there was much basic similarity. While railroads, telegraphs, newspapers, and nearly all, indeed, of the modern devices existed in 1900, they very slightly touched or affected the man living several miles from the railroad, having no telephone, receiving mail once or twice a week, insulated from main currents of modern life by bad roads and clumsy means of locomotion, hemmed in mentally by narrow loyalties and petty conservatisms, by all the totems and tabus, the illusions and prejudices, the parochial outlook resulting from very limited and infrequent contacts with his fellowmen and no contacts with the broad world. The Egyptian farmer at the time of Rameses, the Chinese farmer long before Confucius, the Roman farmer of the time of Cato— men of all times and all races had lived a life essentially like the life of this Maryland farmer of 1900 despite the improved domestic and agricultural equipment he possessed. All of these men were inexorably cut off from the common life of the race. Even centralized authority touched them but tenuously and occasionally and the

1

various seats of centralized authority were, in themselves, but tenuously connected. The vital energies of the race coursed very slowly. Developing and integrated thought and understanding—the blood stream of human society—pulsed very slowly.

But few and far between are the men of any type or kind so detached from their fellowmen in the year 1930 as most men were in 1900. Millions of motor cars, tens of thousands of miles of good roads, radio sets, rural telephones—these are only the more obvious aspects of some of the profound changes which have resulted in two decades. For the cumulative result of all the changes has been to break through all the age-old isolations, insulations, and parochialisms; all the age-old methods of doing things, many of the age-old ways of regarding things. The world seats of centralized authority are more closely connected than any neighboring medieval towns once were. The devices for tying all men instantly and continuously to the seats of centralized authority are intricate and complete. Thought—the blood stream of society—pulses almost feverishly throughout the world, reaching men of all types. The typical farmer anywhere in America in 1930 is living a life radically different in almost ever essential detail from the kind of life lived down the centuries. Not only is he a new type of man living a very different type of life but in his head are new kinds of ideas, inexorably different totems and prejudices. He has, necessarily, a broader outlook. And city dwellers anywhere in the civilized world are kept continuously informed by newspapers, magazines, motion pictures, radio, and numerous other devices, concerning everything which takes place on earth.

We are not here, however, particularly concerned with farmers, city dwellers, or any other special social types as such. We are, rather, concerned with indicating the great and sudden world changes which have come about by showing how profoundly they have affected even remote farmers—that portion of the population which has always changed less rapidly than any other. That the changes have been completely desirable is not to be contended. Too many of the changes represent the swift cutting away of minds from century-old moorings and have brought individual perplexity and unhappiness as well as social instability. But this, also, is apart from our main theme that the progress of the modern world and the combined result of scientific knowledge and ensuing political, social, economic, scientific, and industrial developments in the first quarter of the Twentieth Century represent world changes so complete and so profound that neither man's understanding nor all of the existing agencies and institutions inherited from the past could change rapidly enough to deal with them.

In the material realm, of course, the general interplay of forces brought quick adaptation. Systems of government inherited from the past were overthrown by force and new systems substituted. We have watched this development since the latter half of the Eighteenth Century. We have seen how the evolving knowledge, systematically recorded by Diderot and his fellows, helped precipitate the French Revolution which the rise of industrial wealth, resulting from scientific progress, also influenced. The slender hold of early British imperialism could not maintain enduring grasp on the American colonies. The American Revolution also clearly represents the blasting away of much of the past because of the pressure of radically new conditions and the attempt to found a new political regime by thought and plan.

The application of scientific knowledge to industry made quickly and clearly obvious the inefficiency of inherited, empirical ways of doing things. Methods of industrial production and distribution were quickly abandoned, new and more efficient methods came rapidly into being. Old devices were rapidly replaced by new devices. In medicine and surgery the empirical procedure of the past was rapidly supplanted by new scientific technique. In every human relation where the physical being or the primary economic and political relations of men were concerned, the structures so slowly built up by endless past generations were ruthlessly overthrown or destroyed and new agencies and instrumentalities, based on new knowledge and designed to serve new requirements, were brought rapidly into being. In all the aspects and relations of human life not so closely touching the physical being or the primary relations of men; in all the more involved social and cultural human relations where vision was more difficult or obscurantism easier, recognition of changed world conditions has been so slow that little or no adaptation or change has been made.

The development of democratic institutions has been coincident with the development of modern, exact, scientific knowledge and has possibly served, despite the many benefits it has brought to mankind, to delay in certain instances the more rapid, common recognition of changed world conditions. Rural legislators who impede or obstruct the protecting of the water supply of a great modern city because they cannot be brought to understand or believe in the " germ theory " or to understand the danger of pathogenic micro-organisms afford one example while the legislatures of Tennessee and other states which prohibit the teaching of evolution, afford another. We are afforded, indeed, daily instances of the startling juxtaposition of ultra-modern knowledge and medieval ignorance which cry for the pen of a modern Voltaire. In the State of Illinois, for example, a radio broadcasting station has been used in a systematic attempt to deny that the earth is round and to prove that it is flat. Hygienic measures, attempts at large scale preventive medicine, the diffusion of knowledge concerning great sweeping new scientific discoveries and theories, meet not only the passive resistance of unorganized ignorance but the active resistance of organized forces motivated by superstition and hostility to change. On a larger scale we note the inexorable evolution of many international activities developing apace, even, occasionally, despite the interference of temporary political power. The learned bodies of all nations exchange records of their work and their discoveries and these records are distributed despite world war or revolution. The mail system is now completely internationalized. Radio and cables function internationally. Several American millionaires have given huge sums for international medical or hygienic work which functions without thought of the arbitrary political divisions of mankind.

The production and distribution of basic essentials, particularly raw materials, is now on a world-wide basis. Geologists have charted the world-wide distribution of coal, petroleum, gold, and other natural resources and world-wide industrial and financial organizations are exploiting these resources. Capital is very obviously internationalized. The rapid motion afforded by modern means of intercommunication is affecting, beyond question, the ethnic situation by the constant intermarriage of people of all races to a degree never before imagined. There are numerous unofficial and voluntary international groups of people interested in various subjects or dedicated to various purposes.

Despite the fact that probably 99% of the race is still intellectually unaware of the new scientific knowledge and of the new conditions, there can be no doubt that this knowledge and these conditions have already laid enduring world-wide foundations and are relentlessly affecting the lives of every human being now on earth. No change has come about save as the inevitable result of new, exact knowlege. Even the great feat which helped make the modern world—the discovery of the Western Hemisphere—came because the superstitious fear of the vast unknown spaces of the Western Atlantic was partly allayed by the ambitions, purposes, and ideas which Columbus and his fellows gained only from the correlation of contemporary astronomical and geographical knowledge and speculation.

To no generation of men should the potency of ideas be more clearly obvious than to this generation for we have seen the rapid changes affecting all the lives of men resulting from the ideas of scientists and inventors. We have watched the rapid world-wide development and use of the airplane from the first days when the Wright brothers flew their crude plane at Kitty Hawk on the Atlantic Coast. We have watched the incredibly rapid development, and the ensuing unimaginable influence, of radio from the days when Marconi was first experimenting with " wireless telegraphy." We have seen the ideas of Pasteur change the whole science of medicine. We have watched the almost daily development of startling new devices such as those bringing pictures by radio instantaneously across the oceans. We have seen, indeed, so much that we have lost that clear and simple vision of life which the more admirable and capable members of ruling and governing classes in simpler and more slowly moving ages and generations were able to maintain.

It is obvious to the chief of a savage village that the children must be taught the essentials of what the tribe, as a whole, has learned and thus be fitted to carry on the arts which the tribe has evolved. It is obvious to the chiefs and warriors that the male children must be taught to use the accustomed weapons, must be taught all the simple arts and sciences evolved by the tribe in its successful endeavors at survival and adaption to its environment. It is obvious to the squaws of the tribe that the female children must be taught the women's arts and sciences— to weave and cook, make pottery and perform all the endless tasks essential to the survival and evolution of the tribe. In a wider and more elaborate field it was obvious to the learned men of every semi-civilized or civilized country in earlier times that the youth of the nation or, at least, the youth of the ruling classes must be given an integrated understanding of the basic essentials of that knowledge upon which the civilization, the happiness, prosperity, and development of the nation or of the ruling class of the nation depended. Whenever this obvious necessity was not realized or when it was not possible to fulfil the responsibility, the savage tribe, the civilization, or the ruling class retrogressed or passed from the scene. History offers many instances of the degeneration of tribes or nations or classes due to the fact that the achievements of the past were enjoyed without being understood and the integrated and synthesized knowledge, from which power and growth had resulted, were not passed on. No analogous complete retrogression is possible under modern conditions. The distribution of scientific knowledge is now world-wide. Knowledge is possessed by many diverse races of men living remote from each other and influenced by endlessly diverse environmental, political, economic, and social conditions.

The apparent failure of the contemporary world to evolve more efficient agencies and instrumentalities for the scientific diffusion of the knowledge which has resulted from scientific method is, moreover, temporary. So much has happened within so brief a space of time; so many new ideas, devices, wars, revolutions, rapid movements have come within one short generation that confusion temporarily has been caused. The coordinating activities of Diderot have been maintained. There is a world-wide coordination of the knowledge we have gained, a world-wide coordination of the learned men who test and integrate that knowledge. There has not been as yet, however, any apparent recognition that the primary new necessity confronting a world in possession of such new knowledge is to plan and perfect agencies for diffusing it to the very largest possible number of human beings able to understand it and, understanding it, to combine into forces which can reshape the world in accordance with it and with the instinctive aspirations and desires of men everywhere.

Despite the fact of ignorant legislatures seeking to prevent by legal coercion the teaching of theories upon which most of modern science is based, despite all the endless glaring instances which every newspaper hourly brings us of the tragic gulf between the knowledge which the vanguard of the race possesses and the unnecessary ignorance which causes such vast portions of the race to endure unnecessary miseries and difficulties, three unquestionably positive and inspiriting aspects of the contemporary situation loom before us. We have the knowledge. And, relentlessly and without delay, as well as without general conscious purpose, the knowledge has changed the physical and the political aspects of the world. Moreover, we have great endowments with vast resources charged with the responsibility of " advancing and diffusing knowledge." This is primary and infinitely encouraging. For it leaves as a relatively less difficult, although more complicated, tedious, and lengthy task the problem confronting us from this time on: the making obvious the necessity for the blasting away of ignorance, superstition, totem and tabu and the replacing of these barbarous survivals with reason and knowledge; the evolution of articulate and conscious bodies devoted to this task and the perfecting of agencies and instrumentalities to accomplish it.

Because the diffusion of science implies the necessity for dealing with the most intangible and imponderable phenomena known—the emotions and thoughts of men—and because " knowledge " and " science " are abstractions which we must under no circumstances objectify and treat as if they were concrete realities, the problem appears difficult or impossible. Yet the task involves only the performing under modern conditions of the exact function which the tribal chiefs and squaws performed in teaching their crude arts and sciences to the youths and maidens who would become the tribe. It involves nothing more than an attempt to attain the maximum social utilization of our modern scientific knowledge *as a whole*. It cannot attain maximum social utilization or bring happiness and well-being to the individual or to the race if the pernicious doctrine of specialization is not relentlessly excised. Many of the most painful aberrations of modern civilization have unquestionably been caused because men, powerful and highly placed, had been given specialized instruction leaving them ignorant of the relation of the few or many facts they knew to the complete knowledge of their time.

An international organization desiring to secure and distribute any natural resource like coal or oil or timber or food would proceed in an orderly manner and by scientific method. It would chart all the available supplies. It would chart all the possible methods of production and distribution. It would proceed by orderly and systematic plans in the pursuance of its object and purpose. Those who would set themselves the task of securing the maximum social utilization of modern knowledge by its most scientific diffusion must proceed in a manner relatively analogous. Out of the vast area of all that men have believed, we must chart an area of relative certainty. An international commission of humanist scholars of worldwide reputation can integrate the basic essentials of the most important things which the race as a whole now knows. Full understanding of the new knowledge as a whole and of its relation to individual and social happiness and progress must be sought. There must be careful study of the nature of the philosophies of the past concerning the diffusion of knowledge, and of the instrumentalities for the diffusion of knowledge built up and now existing as a result of these philosophies; there must be study of the evolving agencies for the diffusion of the new scientific knowledge which recent generations have secured.

Despite the fact of the conscious revolutionary changes in political and industrial organizations there has been no far-reaching, *planned* change in educational philosophy, material, and technique. In an ultra-modern world our instrumentalities of instruction are largely empirical and inherited from a past totally different from today. The new era upon which we are obviously entering, the new world which will, unquestionably, come into being requires for its more rapid evolution the conscious and deliberate formulation of new philosophy, material, and technique of education clearly designed to bring to as large a proportion of the eighteen hundred million human beings on earth as possible the clearest possible understanding of the most vital aspects of the modern knowledge which has not only changed the world and all the relations of men but has given to truly educated men throughout the world an outlook upon life and existence which men of no previous generation ever had.

II

THE KNOWLEDGE OF THE PAST AND THE AGENCIES FOR DIFFUSING IT

In one sense, of course, knowledge is the product of five hundred thousand years of human life. Indeed, it might be contended that the roots of knowledge antedate the evolution of man. There are endless theories of knowledge but the word can, for our immediate practical purpose, surely be properly applied to the articulated experience of the race; the ever increasing awareness concerning every aspect of the human and non-human universe gained by the special senses and the general abstracting, sensory organization of man coordinated by his cerebral cortex and finding expression through symbols forming a language of some type. We have no evidence that knowledge was ever other than empirical down all the evolutionary process until a few thousand years ago. To the multitudinous generations of primitive man intelligence was merely one of the forms of human behavior—a form of behavior not possessed to the same extent by the other types of life and giving a possibility of dominion over them. It has been pointed out by Alfred Korzybski in his " The Manhood of Humanity " that the essential characteristic of the human class of life is its capacity to " bind time " by transmitting experience.

We distinguish between " science " and " art "; between " knowing," and " making " or " doing " as a consequence

13

of knowing; but by primitive man no such distinction was made. To him art and science were one and knowledge was probably transmitted and augmented rather by the evolving artifacts representing evolving and accumulated experience and adaptation than by explanations in words. So far as we know at present the first conscious abstractions, the first understanding or attempt at understanding of synthesized or integrated experience, expressed in symbols, cannot date back much more than twenty or thirty thousand years. We need not here be concerned with the more involved aspects of this development but may concern ourselves with the obvious existence of consciously recorded, non-empirical knowledge by the scholars of China, Assyria, Egypt, and possibly Mexico and Peru, five or six thousand years ago. In this recorded, non-empirical, knowledge there was obvious the essential quality of what today we call " science "—the attempt to integrate exactly obtained and recorded experience and to crystallize from these data abstractions or generalizations entirely objective and susceptible of communication, demonstration, and verification.

Leisure and the political or other social power permitting leisure, were obviously essential for the men concerned in any way with these new types of highly evolved human behavior or activity so that, through thousands of years of the historical period, we find knowledge in this sense developed by, possessed by, and transmitted by only the small groups protected either by the political power of kings and other potentates or by the religious power of priests. All the conditions of semi-civilized society mili-

tated, moreover, against the rapid augmenting or the widespread diffusion of the tiny body of non-empirical knowledge which was being evolved and developed only very
slowly over generations and centuries by groups so small,
in relation to the general population of the earth, as to
appear negligible; to constitute, indeed, a relation as obscure and unperceived as the relationship which the human
cerebral cortex bears to the human organism as a whole.
Inevitably, magic, superstition, political and ecclesiastical
policy and numerous other factors would affect and influence this small body of knowledge which—as a matter
of historical fact—developed largely only in the mathematical and astronomical fields. The facilities for recording this knowledge, hopelessly intertwined and infiltrated
with mysticism, speculation, and chicanery, were, moreover,
extremely meager. It may well be believed that even three
or four thousand years ago only a few score of men in
any world-wide generation would have information concerning even the mere existence of the small body of
recorded mathematical and astronomical knowledge. And
we have seen how, even so, this knowledge was almost
completely lost—only the practical devices which it had
brought into being remaining to influence human life—
during the so-called "Dark Ages." Modern historians are
continuously affording us new, exact information about all
this evolution and development as a phase of history.

For our present purposes it is essential only to recognize
a few basic facts concerning the knowledge of the past
twenty-five centuries. Among these may be stressed the
general Aristotelian influence which, in the Occident,

caused the development of exact knowledge always to suffer the handicap of *a priori* speculation; the continuing difficulties of record and transmission until the development of printing; the obvious political and social advantages to ruling or exploiting classes in maintaining a firm monopoly of existing and evolving knowledge; and, finally, the meagerness of this knowledge which rendered it obviously possible for any member of the upper reaches of a political or ecclesiastical hierarchy with sufficient personal desire and leisure to gain rounded understanding of it in his lifetime.

Francis Bacon, the man whose intellectual audacity and thoughts have so largely served to create our modern world, boasted in his youth in Aristotle's phrase: " I take all knowledge to be my province." And it was an ambition seemingly possible of consummation. Even so late as 1585 there probably appeared no insuperable obstacle to the endeavor of an ardent young intelligence to obtain understanding of all the knowledge which mankind as a whole was in conscious and articulate possession.

Bacon's desire voiced the implicit, and often explicit, educational philosophy of all the ancient and medieval world—pansophism—the desire to know everything; the attempt of the dynamic human intelligence to expand harmoniously out into a vast universe of recorded experience but not yet conscious of the fact that this universe of so-called " knowledge" was shifting, unstable, and unreal, because metaphysics, speculation, unverified history, and unverifiable theory, magic, illusion, superstition, and prejudice made claims to being part of " knowledge " and, using words and symbols to find expression, rendered such conquest of knowledge as Bacon planned a task as fore-doomed to failure as the conquest of shadows in a confused dream. It is because the Bacon of later years doubt-less came to recognize this fact; came to see the dark and ugly, the unreal and sinister shades which had intruded themselves into the idea of " knowledge," that he wrote the *Novum Organum* and helped bring into being this relatively sunlit modern world of airplanes, radio; of exact knowledge giving such control as was never dreamed while it becomes so intricately detailed and voluminous as was hardly to be imagined.

The geographical knowledge of Herodotus, Strabo, Pliny, and others was no more a confused mixture of geographical truth and vivid unreal imaginings than all the general learning of the world until very recent times. Even in the story of " Sinbad the Sailor " in the Arabian Nights, and in the writings of Sir John Mandeville, there was a sub-stratum of realistic, geographic facts, but to all the scholars of the past just where fact ended and fancy began was never clear. In modern histories of the developing explorations of the planet, we can watch the unrolling of the clouds, the increasing range of geographic information of the ancient world, the increasing precision of the maps, the constant exclusion from serious consideration of fanci-ful tales of Prester John and men with eyes between their shoulders. The unrolling of the clouds and the emergence of widespread recognition of the difference between the verifiable generalizations of science and all the fancies with which ancient learning was cluttered up is almost exactly analogous. It requires definite effort of the imagination for a man under the influence of modern thought and conditions to visualize convincingly the medieval world out of which modernity has sprung with such prodigious speed and reached heights so great. It was a world which had seen a thousand years of relatively fruitless efforts to re-establish effective centralized authority; a world of muddy roads and fighting men, of moats and walled cities; a world where, even when the Renaissance caused new appreciation of a sunlit past and a possible sunlit future, interest was centered largely on the aesthetic and cultural aspects of human betterment without clear recognition of the importance of the scientific and technical aspects which must, perforce, support them.

The impulse in the human mind to expand is as inevitable as the impulse in a seed to expand and grow. The periods in which there seems apparent intellectual inertia or apathy will probably be understood by a better informed future as periods hemmed in by dogmas and creeds, totems and tabus, which rendered the expanding of individual intelligence hopeless and resulted in the intellectual stagnation of which scholasticism is an example. There has probably never been any period since the fact of the mere existence of knowledge was known to any part of the general population that numerous men have not eagerly desired to know, to grow intellectually, and to be willing to make any sacrifice or effort to acquire the knowledge of which they had dimly heard. Underneath all the more obvious aspects of the ancient and medieval universities and monasteries this eternal human characteristic must certainly have manifested itself. And, despite the close monopoly of learning; despite the difficulties of record and transmission, the absolute number of men more or less well-informed was, because of the increase of population and the general development of pragmatic culture, ever increasing throughout the ancient and medieval worlds. The actual progression of the race, the general rise of culture in the sense of control of environment and perfecting of devices may probably have been due to the mere unformulated and unrecorded development of essential human capacities in dealing with the non-human aspects of existence. The knowledge which an ever increasing absolute number of priests, clerks, attorneys, philosophers, and teachers possessed may have temporarily remained relatively static because of the Aristotelian influence but there was ever increasing dim recognition of the existence of evolving knowledge and of the necessity for recording and transmitting it.

The surprising fact has recently been pointed out that the whole idea of progress is a very recent development. The rapid rise of the modern world might, in one sense, be considered as being partly due to man's sudden recognition of his power to adapt himself to, or to control his environment. It is only when this thought comes that knowledge is rendered dynamic. And this thought can only come when exact scientific knowledge has expressed itself in devices which show to the meanest intelligence man's power of changing his environment according to his needs and desires. Alexander, Caesar, Napoleon—all the conquerors, all the more dynamic peoples—clearly saw their power to subjugate and to control other men or nations. The subtler idea of man's power to control his environment is distinctly an evolution of very recent years. The relative stagnation of the past, the enormous energy and movement of the present may well be considered as being, respectively, due to former failure to make the generalizations of progress and possible control and our present clear-cut universal accentuation of such generalizations. It will be remembered that it was not until Huxley's famous essay on " Evolution and Ethics " that the clear-cut expression of man's power to combat the cosmic process was made. None of this, of course, was clear to the ambitious men of earlier periods eager for mental development. Yet, no matter what blind alleys and morasses into which their minds may finally have been led, the thousands of men who trudged hundreds and thousands of miles over dangerous roads—chancing physical violence, brigandage and dangerous despoiling of every type—surely desired to gain understanding. One of the basic impulses which motivated them in going to famed monasteries and universities was surely to acquire knowledge; to fulfil their essential human destiny, to keep alive, to augment and to transmit the crystallized experience of the race. And deep beneath the befuddlement caused by theology and speculation, there was somewhere in all these monasteries and universities a confused recognition of this necessity. The will-to-power, the desire for personal preferment are the obvious motives which casual observations show, but intertwined with these, unquestionably, were deeper and purer impulses.

From the time of the early chieftain and squaw who had taught the youths and maidens the simple arts and sciences as a whole, knowledge had grown vast and confusing. And because it gave political, economic, and social power and because it had to be expressed in words, those charged with diffusing it had grown perplexed and befuddled. For long centuries it almost appeared that the essential purpose had been hopelessly obscured. The retrograde centuries from the fall of the Roman Empire to the period of the Renaissance indicate what happens to mankind when this vision is lost or this essential task only poorly accomplished. For a thousand years endless interrelated forces caused the isolation, insulation, and immuring of what confused learning or what desire for learning still remained or evolved. And numerous practical and immediate conditions resulted from this.

In the ancient world, in China and in Rome the centers of power had direct, immediate, and constant relations with all those men of learning whose knowledge could be made of practical use in the maintaining or augmenting of power. Ching directly controlled the engineers who planned and directed the construction of the great wall of China. The engineers and artists of ancient Egypt were under the direct and immediate control of the emperors who used their skill to construct pyramids, great tombs, colossal rock carvings, and other grandiose works. The Roman oligarchy had direct and immediate contact with the trained men who constructed the aqueducts, the great Roman roads and walls, triremes, and all the devices which permitted Roman power to spread ever farther afield.

In the political and social aspects it was a result of the lack of centralized authority in Medieval Europe that permitted the development of the powerful monasteries— insulated centers where learning was not only immured but where human intelligence was temporarily freed from immediate, pressing necessity or coercive power and permitted to branch out either into futile speculation or into the relatively disinterested research akin to the pure science of today of which it was the forerunner. Mere literacy became a somewhat mysterious badge of social superiority, and there was the evolution of very highly specialized social types becoming ever more greatly differentiated from the great mases of the population while having less definite and direct relations with political authority.

The eternal struggle between political and ecclesiastical authority also assumed somewhat new aspects. Many a medieval princeling felt the intellectual concerns of monastically reared scholars to be as remote from the realities in which he was most interested, and with which he was most concerned, as any ignorant serf might have felt them. The identity of state and ecclesiastical power which the ancient world had known had disappeared. In large measure probably any knowledge possessed by the priesthood of the Roman Empire was directly at the service of the political oligarchy with which the priesthood was closely associated. For that reason Roman noblemen could buy learned Greeks as slaves largely to entertain or delight them with cultural achievements rather than to help them maintain or augment realistic power. The tutorship of the young Alexander by Aristotle represented a use of learned men by ruling classes which was largely to diminish during the next fifteen hundred years as divergence between political and ecclesiastical power developed. It was inevitable, therefore, that there should eventually evolve a new type of specialized practical instruction designed to teach the sons of the powerful the technique of government and the maintainence of power. Only incidently could this instruction be gained from monastically trained men who had the implicit idea of pansophism. Machiavelli's "The Prince" is one indication of the evolution of a new type of instructor seeking to gain power and place—not motivated either by desire for knowledge or to entrench the power of the church but to gain the favor of rulers by specializing in the study of political realities. This motivation marks the development, under feudalism, of complicated educational institutions which our modern world inherited. The English "Public Schools" and the great English universities have historically born something of this relationship to the English governing classes although, by slow and inevitable evolution, the universities at least, naturally acquired a broader social purpose and outlook. We see, however, in the evolution of the monasteries and of the general type of educational institution represented by the English "Public Schools," two distinct kinds of educational institutions evolved by the past to diffuse knowledge. The completely rationalized folkways, moreover, inevitably evolved other types of educational institutions. With the continuing development of all arts and handicrafts; with the ever more complicated structures and devices men learned to perfect, schools for the training of artists, artisans, and craftsmen of many kinds necessarily developed. With the passage of generations these schools assumed increasingly complex forms.

The increase of wealth, even before the Renaissance, had developed ever more numerous classes desirous of having their children given the opportunity to attain high social place. The monasteries responded to this social need by initiating schools. Later these schools were detached from the monasteries although, for many centuries probably, they were conducted by the "clerks" originally educated under monastic influence. It is readily to be seen that the ruling classes would watch with suspicious eyes the development of such a phenomenon and that those entrusted with the education of the children of the non-ruling classes would stress the necessity for remaining content with that station in life in which it had pleased Providence to place one. Moreover, the instruction given in such schools would necessarily refrain from dealing realistically with contemporary social, political, and economic realities and, being inhibited in many other ways by the necessity for discretion, would afford instruction very definitely lacking in broad vision or vigor. It may surely be declared that, so long as it was in any way possible to obstruct it, all the diverse classes seeking control of society were united in their fear of the dissemination of knowledge to the general population and worked to keep the mass of the population in ignorance, fearing selfishly for themselves, and also to some degree fearing for society in general, if the force of realistic thought was permitted to infiltrate society as a whole. This age-old attitude is made strikingly clear in some of the letters written by Catherine the Great.

In broad outline such were the more essential aspects of the knowledge of the past and some of the outstanding agencies evolved to diffuse it when numerous unexpected forces evolved to change man's outlook everywhere and to bring into being the situation with which the contemporary world of today is familiar.

Four forces generally affecting the knowledge of the past and the agencies for diffusing it have been stressed herein. The influence of Bacon and others served to begin to disrupt the first of these influences—the handicap placed upon the development of exact knowledge by the Aristotelian tradition. The perfecting of the printing press and the slow evolution of more efficient methods of intercommunication served to disrupt the age-old difficulty of record and transmission of evolving, exact knowledge. The industrial revolution, the growth of widespread elements of the population possessing increasing wealth and power, and the inevitable rapid influence of the press rendered increasingly difficult or impossible the age-old policy of the ruling classes to monopolize knowledge.

The application of the Baconian ideas, the influence of the Renaissance, the tremendous stimulation of the European intelligence due to the discovery of the Western Hemisphere, the rapid and enormous increase of literacy with the consequent increase of investigation and research, caused detailed, exact knowledge to augment so greatly that it became increasingly obvious that no one man in a lifetime—no matter what his advantages—could hope to learn in complete detail all that mankind as a whole had come to know. This development, however, is much more recent than the other three and has, to great degree, influenced the minds only of the past three or four generations.

The chief underlying purpose of this book is to make clear the crystallization within the general body of human experience as a whole of that area of relative certainty—the modern scientific knowledge which does not deal with speculation or mysticism and is relatively exact, verifiable, demonstrable, and communicable.

What is new in the lives of modern men has resulted because a few men throughout the world have developed and applied such scientific knowledge. One of the characteristics of this knowledge which has not been sufficiently recognized is the fact that it is dynamic not only through what it teaches us, but even more so by what it shows us to be false in many of the primary sense experiences of the race and in most of the accumulated beliefs of the race. These beliefs—part of the folkways—inhibit mental growth. It is because this mental growth of the great masses of the population is thus inhibited, while a small portion of the race possesses and applies the new knowledge, that most of the perplexing social aberrations of today are possible. It is for this reason that radio broadcasting stations are used by ignorant bigots to seek to promulgate the idea that the world is flat. It is for this reason that rural legislatures prohibit the teaching of the evolutionary theory. It is for this reason that such vast sections of the literate but ignorant masses of modern democracies are left to the tender mercies of pandering demagogues, journalists, and ecclesiastics. The little understanding which the individuals constituting these populations possess of scientific knowledge is as hopelessly intermixed with all the ancient illusions and superstitutions as the records of the ancient explorers are hopelessly intermixed with childish tales. Lack of understanding of the knowledge possessed by those in the vanguard of contemporary civilization causes constant and often effective obstruction to endless movements seeking only the common good of mankind. Such obstruction cannot be finally prevented until the knowledge possessed by the vanguard is infiltrated as far as possible through society as a whole. It is desired to intimate that, slowly out of the inarticulate, empirical knowledge of the race over hundreds of thousands of years, there has been crystallized a small but tremendously dynamic body of relatively exact knowledge which constitutes now the most valuable and potent possession of mankind.

The attempt has been made to indicate the slow evolution of this body of knowledge in the past and to indicate that the past—never having formulated the idea of progress nor having properly conceived the potency of exact knowledge—failed to evolve effective instrumentalities for the widespread diffusion even of that part of it of which it was in possession. In any case these instrumentalities must have been meager and relatively inefficient as all our present machinery for widespread, universal intercommunication is a by-product of the development of exact science.

But let us remember that the ancient world attempted to see what knowledge it actually had as a rounded and integrated whole. For there is reason to believe that one of the forces most disastrously militating against the widespread diffusion of exact knowledge today is the sinister educational philosophy of specialization which so completely influenced the latter portion of the Nineteenth Century. This statement of our underlying purpose is made to throw into proper perspective the first three of the new factors which have been pointed out as helping to disrupt the general situation existing until the time of the Renaissance.

The influence of Bacon helped to send men afresh and with fewer preconceptions to the careful investigation of phenomena, to purge learning of confusing irrelevances, and to help the crystallization of which we have spoken.

When Mr. Stuart Chase estimates that sixty thousand words per capita, per day, are printed in the United States of America; when we know that five hundred thousand copies of the works of George Brandes have been sold in Denmark alone; when newspapers of a million circulation and weekly magazines of a circulation of two million and a half are familiar phenomena, we obviously need not linger to discuss the far-reaching influence of the invention of printing and the development of all the devices of inter-communication.

Diderot, almost always impecunious and coming from a class in no way closely allied to the governing class of France; Diderot, sitting at a desk for twenty years editing the Encyclopedia which was to help precipitate the French

Revolution, is, likewise, a sufficient symbol of the development of ever larger, less closely cohered groups possessing knowledge despite all the efforts of rulers, and valiantly laboring to give it to mankind at large.

All three of these influences are too obvious to need much comment. But the fourth aspect of the knowledge of the Pagan and Medieval worlds, the meagerness which rendered it obviously possible for any member of the upper reaches of a political or ecclesiastical hierarchy with sufficient personal desire and leisure to gain rounded understanding of it in a lifetime which was disrupted by the rapid and enormous increase in literacy with the consequent increase of investigation and research causing detailed exact knowledge to augment so greatly that it became increasingly obvious that no one man in a lifetime—no matter what his advantages—could hope to learn in complete detail all that mankind as a whole had come to know, cannot be so easily dismissed.

Here, in fact, is possibly the underlying cause of all the unending criticisms which, throughout the world, are being made of contemporary educational philosophy, material, and technique. For with the increasing complexity of detailed knowledge—the increasing difficulty of recording, save in highly abstract, intricate, and technical form, the remote, obscure, and intangible phenomena with which research dealt—scientific knowledge inevitably lost humanist quality. It came, to the average uneducated, or partly educated man, to appear altogether too far remote from his own life, happiness, and ways of being.

Within five generations the world gained a larger body

of exact knowledge than all preceding generations had built up. This knowledge came, moreover, during singularly restless and hectic times, the changing of vast social systems, the colonizing of vast new continents, the questioning of inherited authority of every sort, the emerging from the relatively static social conditions which thousands of years had known, of highly individualistic, almost anarchic, social conditions so dynamic as to cause an almost universal sense of perplexity and confusion. All professions and industries developed with unheardof rapidity. Demand existed everywhere for highly specialized and detailed technical knowledge in many very minor fields of the great provinces of knowledge as a whole. Quickly there was the evolution of technical schools of endless diverse sorts to fit men for these new highly specialized functions.

To so-called " practical " men, especially to the new type of aggressively individualistic entrepreneur evolved by modern capitalist industrialism, and to the alert and vigorous men who were building the railroads and industries of America, the value and importance of disinterested concern with pure science from which, alone, the very devices which gave them their power had resulted, were obscured and lost. All the more dynamic social types evolving with only tenuous ties to new types of elected, temporary, democratic political authority were, indeed, contemptuous of any knowledge or activities which did not bring quick and immediate practical results.

To the chieftain of the primitive tribe the value and importance of the knowledge of the tribe as a whole was obvious. To the potentates and monks of the Middle Ages, to the young Francis Bacon, the value and importance of knowledge as a whole were still confusedly obvious. To the powerful industrial classes of modern societies understanding of the value and importance of integrated knowledge—knowledge as a whole—appears to be completely lost. Yet the control and influence of these men in the newer modern democracies have been very great. *Their* desires and *their* opinions have profoundly influenced governments and enormously influenced the colleges and universities frequently endowed or supported by them. The colleges and universities, in fact the whole educational system from kindergarten to post-graduate school, has felt the pressure of endless professions and industries clamoring for men who, however broadly ignorant, had been trained to render efficient service in some highly specialized field. The effect of this continuing pressure upon modern educational philosophy has, inevitably, been to cause the rationalization of a social activity thus influenced.

The closing years of the Nineteenth Century heard widely proclaimed the cheap slogan: " It is infinitely better for a man to know everything about something than to know something about everything," a slogan quoted merely because it represents a concentrated expression of the prevalent viewpoint in educational circles. Such a program is, of course, unworthy of comment. It is clearly impossible for a man to know something about everything. It is even more clearly impossible for him to know everything about something.[1]

The truly liberating and dynamic effect of detailed, scientific understanding is not released until the true relationship of this detailed understanding to scientific knowledge as a whole is clearly seen.

The slogan may have alleviated the deeper doubts of men forced to produce ignorant technicians but it could not, of course, be developed into an enduring educational philosophy nor be accepted by socially conscious scholars imbued with a true tradition of learning. Yet it has profoundly influenced the educational practice of the past few generations and is very obviously and markedly in being today. It has resulted in the development of learned institutions filled with savants so lost in the study of one highly specialized branch of knowledge as to retain with difficulty, if at all, broad understanding of knowledge as a whole and of the direct, immediate, and unceasing relationship of knowledge as a whole to individual, social, and race well-being. It has resulted in educational institutions of all types designed to fill minds with a mass of uncoordinated facts, preliminary to highly detailed specialized instruction of some sort. It has resulted in the development of great masses of men filling specialized social functions, informed concerning some one branch or field of knowledge but realizing only very slightly, if at all, the true relationship of their knowledge and activities to the history, knowledge, and life of the race as a whole.

For the ideal social systems imagined by ancient philosophers, where a small ruling class was always assumed to represent mankind while resting on the shoulders of an unconsidered slave population, this development of lopsided minds highly trained in one field and almost completely ignorant of life, history, and knowledge as wholes might be ideally fitted.

But the democratic ideal and tradition too greatly suffuses the contemporary world. And the benefits to any governing class of having highly specialized and relatively ignorant workers have been more than overcome by the obvious social disadvantages to all classes through the existence of such a condition. The few men in the vanguard of the modern world who have integrated understanding of exact knowledge are, moreover, widely separated from the apparently befuddled political rulers who have little need, as well as little understanding, of rounded scientific knowledge; who are motivated by traditions of the past and only dimly, if at all, conscious of the existence of the new age which has come so rapidly into being.

Strangely enough specialization in education developed most markedly in the United States of America where the democratic ideal might have been assumed to prevent such a development. The reasons are obvious. After the success of the American Revolution endless tasks confronted the new nation. Social stability had to be gained and the new institutions of government tested and made to work. The eyes of all men were on the vast territories to the West. Despite all the teeming vitality and energy too many problems clamored for attention for all of them to receive effective treatment.

The relationship of universal literacy and of an informed electorate to any enduring democratic state was very obvious. The fact that most contemporary educational institutions were the product of past ages *without the democratic ideal* was not obvious. Public schools were initiated throughout the country and an almost unprecedented portion of the population was rendered literate. It may, however, be contended that the general educational system represented nothing entirely new, was the product of no clearly thought-out philosophy but was rather a mélange of all the various educational institutions, devices, and purposes evolved up to that time. A conscious revolution had taken place in the political realm. Inherited political institutions had been disrupted and new, *planned* systems brought into being. Had it been possible to disrupt in a similar fashion the inherited educational philosophy and institutions and to replace them by new institutions planned to serve the obvious new purposes, the whole development of the United States during the past five generations might well have been very different. But this is a belated counsel of perfection. It is the recognition after the event of a necesity which could not have been clear at the time nor to which was it possible to apply the thought and effort which would, obviously, have been necessary. There were, in relation to earlier systems of education, minor and major differences it is true. And the very spirit of the instructors and teachers was certainly different. The deliberate obstructive policies of upper classes no longer inhibited teachers in their efforts to diffuse such realistic knowledge as they possessed.

Until the time of the Civil War the vivid and generous energies awakened by the Revolutionary period unquestionably so affected American educational institutions that, despite the lack of a clearly formulated new educational philosophy, teachers did succeed in infiltrating everywhere an emotional, if not an intellectual, appreciation of many aspects of the new order of things which was coming into being. But this understanding was suffused with a patriotism which tended slightly toward chauvinism. Many powerful new classes were evolving in America. The democratic ideal was beginning to display aspects which had not been foreseen. Just as the monasteries during the "Dark Ages" had retained only a confused sense of the purposes they were seeking, so the American educational system gives evidences of beginning to lose clear vision of the purposes it had been planned to serve. Emotionalism is not enough. There must always be made and retained the clear-cut formulation of a purpose. It is not without reason that less than half of the American citizenry entitled to vote uses its right of suffrage. While there was the rapid development of widespread democratic educational institutions it cannot be successfully contended that the educational institutions of America have represented essentially any vital and effective new articulate instrumentalities for diffusing knowledge. America evolved a complicated and costly educational system beyond anything the world had imagined but this system was new only in size not in its basic nature.

The disadvantages to society of specialized instruction have become so obvious that there is everywhere now apparent a new insistence on the " humanizing of knowledge " as well as a developing recognition of the necessity for the humanist integration of knowledge and for more conscious and deliberate efforts seeking the diffusion of knowledge thus integrated and humanized. It is now being recognized that the man who knows a great deal about some one aspect of science without having broad understanding of the relationship of the knowledge he possesses to the whole body of knowledge is at once an unstable and undependable social element and a man rendered unfit for the happiness and satisfaction in life which even savages find in an environment of which they have rounded, however, limited comprehension.

To geographers and mariners of the early Sixteenth Century the exact shape of Western Europe, of the Mediterranean, the Northern Coast of Africa, and the Red Sea might have been known. But there were still such hazy ideas of the new areas of the world being discovered that there was prevalent general belief in El Dorado, Patagonian giants ten feet high, rich civilizations in the interior of North America, and endless other unjustified beliefs. So a typical man of today may be almost a world specialist in one of the sciences or may be a technician using daily the most detailed information yet, lacking broad understanding of history and of contemporary civilization, be subject to the most surprising prejudices and passions and respond to the crudest appeals based on tribalism, superstitition, and passion.

The World War showed that many famous scientists lost all world vision when their unrationalized prejudices were sufficiently stimulated. There is something obviously at fault when a universally literate population presents the social and political aberrations displayed every day in America, England, and other countries. The perfecting of

printing and engraving has served to permit many great
modern cities to be flooded with newspapers, the implicit
appeal of which is to utterly savage and barbarous characteristics. Millions of people educated under modern
conditions respond vividly to motion pictures which represent appeals which, in earlier periods, would have been
successful only with grossly ignorant sections of the population. There is reason to believe that if our systems of
education had not suffered by the philosophy of specialization and if all children from the beginning of their
instruction were given a broader and better coordinated
outlook these political and social aberrations, these newspapers and motion pictures would not today exist.

In the sense in which we have, throughout, used the
word " knowledge "—the " crystallized and articulated experience of the race; the ever increasing awareness concerning every aspect of the human and non-human universe
gained by the special senses and the general abstracting
sensory organization of man coordinated by his cerebral
cortex and finding expression through symbols forming a
language of some type," it is obvious that every child born
on earth should be given such humanist understanding of
it that it will see the relationship of all that the race has
learned to its own individual growth, happiness, and well-
being.

The impulse which actuated the primitive chief, the
impulse which prompted the young Bacon, are still unquestionably valid no matter how complicated knowledge
has become. And, if it has become obviously impossible
for any man to know all the details of all that the race

has learned, it is certainly *not* impossible for any man to learn the broad basic essentials of all that the race has learned, to have that intellectual vision which will render him immune to the illusions and superstitions which still inhibit the mental development of 99% of the race.

We may conclude by declaring that the knowledge we have inherited from most of the past is not precise and relatively exact, verifiable, communicable, and demonstrable but is intricately intermixed with metaphysics and mysticism. It developed only very clumsy instrumentalities for its widespread diffusion. The chief agencies for diffusing it originated, moveover, in a monastic tradition of isolation and insulation from the common life of man; originated in an age-long close association with potent social forces desirous of keeping the great masses ignorant. But that knowledge *did* implicitly seek integration and *did* seek to maintain humanist value.

The first phase of the development of the more vital knowledge which we inherit from a much more recent past represented the intricate subdivision of this knowledge into arbitrary divisions not well integrated nor kept in close relationship to human desires and needs.

III

THE NEW SCIENTIFIC KNOWLEDGE AND THE EVOLVING AGENCIES FOR DIFFUSING IT

The word "new" means nothing without clear definition and a time coordinate. In one sense all the truly scientific knowledge gained from the time that Francis Bacon wrote:

We entreat men to agree to discard . . . these volatile philosophies which . . . have led experience captive . . . and to approach with humility and veneration to unroll the volume . . . to linger and meditate thereon, and with minds clear from opinion to study it in purity and integrity—

may be designated as "new." For our specific purposes in this volume let us, however, rather somewhat arbitrarily apply the word to the general range of ideas inexorably bringing a new integration of scientific knowledge and breaking ruthlessly through the arbitrary divisions into which scientific knowledge had been separated. The influence of the electronic theory and Einstein's theory of relativity have been most potent in this respect. We may, accordingly, define what we shall call the "new scientific knowledge" as the whole body of scientific generalizations which have been profoundly influenced by these ideas and therefore fix our time coordinate as post-1900.[2] But this is only the central core of the matter. It must be clearly understood that the influence upon modern thought of the cumulative effect of totally new conceptions of evolutionary time, the forcing of man's imagination back millions, tens of millions, even hundreds of millions of years by the advances in astronomical, geographical, paleontological, biological, and even archaeological and historical sciences forms one of the bases of the new scientific knowledge, as does, likewise, the influence of the new, precise knowledge concerning the vastness of interstellar space resulting from common use of the idea of thousands or millions of " light years." In many instances, of course, progress has represented no " new " discoveries but, rather, "new " understanding and correlation of existing knowledge. The evolutionary theory and its influence in causing a new sense of the unity of life must also be taken into consideration. From the fields of mathematics, astronomy, physics, chemistry, biology, geology, paleontology, anthropology, and psychology have come, since 1900, new discoveries and theories, the combined influence of which has been to make clearly obvious to the learned world the inexorable new integration of science which has resulted. And many specific contributions such as those afforded by physical-chemistry, colloidal-chemistry, biology with its theory of tropism and the insistence upon consideration of " the organism as a whole " must not be overlooked.

The era of crass specialization during which the original divisions of scientific knowledge, once understood as having no more real existence than lines of latitude and longitude, were treated as realities, must come to an end.

In 1930 astronomy and physics have been brought so close together that the differentiation between them is difficult to establish. For we see in the hottest gaseous nebulae, the type B stars, the probable existence of basic energy which is just beginning to assume the simple form of hydrogen and helium. We see through the spectroscope, in the next hottest, type A, stars, the definite presence of hydrogen and helium. The next in order, the class F stars, are characterized by an abundance of calcium. They also afford evidence, through the spectroscope, that metals are about to be developed but have not yet arrived. None of this is comprehensible to the astronomer without the knowledge which the modern physicist has afforded him. Nor can the modern physicist find proof for some of his generalizations save from the astronomer.

The new close relations which the past few years have established between physics, chemistry and biology are too obvious to need comment. We now see clearly that just as the physicist in studying the building up of the element looks upon the electron as a unit, so the chemist in dealing with the various possible combinations looks upon the entire atom as a unit, and just so . . . the biologist looks upon the still more complex structure, the cell of protoplasm, as a unit from which biological structures are built up. We begin to see that " life is a phenomenon of energy not yet fully understood " and that " biological laws are but physical and chemical laws operating in those higher planes of their application, the more complex colloid structures." Inevitably these basic conceptions influence the anthropologist, the historian, the psychologist, and the sociologist. History is no longer seen as the mere record of dynasties and nations. We see all the behavior of the various civilized peoples of the past five thousand years (and it must be remembered that, even a century ago, history gave practically no consideration to China, Japan, and the Mayas) against the background of the hundreds of thousands of years of human life which had preceded them, which, in turn, resulted from the general evolutionary process.

No serious attention is paid today to history written without reference to the discoveries of all the modern sciences. The generalizations of the evolving science of sociology are, similarly, valueless unless they are rooted in the work of sciences which, only a generation ago, were considered to deal with " extra-human phenomena." Mathematics and logic have been brought strangely close together in recent years while a world renowned scientist has recently proclaimed that philosophy, henceforth, will assume merely the place of the " history of the sciences." Almost like the breaking down of arbitrary barriers or of national frontiers and the sudden unification of physical areas has been the result of the impact on the learned world of the basic new generalizations of the past two decades. From mathematics to psychiatry the various divisions of scientific knowledge have had their frontiers blasted away and the whole field of modern scientific knowledge made one; made so completely one that the true meaning or position or value of no part of it is clear without broad vision and clear perspective of it as a whole.

Another basic aspect of what we are attempting to define as the " new scientific knowledge " is the complete blasting away of the assumed absolutes and objectified abstractions which have poisoned and inhibited men's

minds and prevented the rapid advance of human understanding through all eras until today. The theory of
relativity—applied to all knowledge and to all human
relations—will increasingly influence the minds of men
from this time on. At the heart of every philosophy and
every religion which mankind has known, there has been
a definite or implicit absolute. Science admits no absolutes.
We have seen in recent years the far-reaching influence
of the disintegration of the age-old conceptions of " absolute time " and " absolute space."

Note that modern thought, emerging from Newton,
does not even refer to " the universe " but to " a universe,"
recognizing with Einstein, that the Newtonian universe
is but one possible explanation of the knowledge we possess. Throughout all modern investigation all absolutism
is rapidly disappearing. And it is noteworthy that even in
the religious and political realms the unperceived influenced of the contemporary *Zeitgeist* is to diminish or
remove all absolutisms.

What has been the first effect of this new inexorable
integration of the basic essentials, the basic generalizations,
of all the divisions of knowledge?

That integration has, as yet, taken place consciously only
in the heads of a relatively few men in the vanguard of
contemporary scientific thought and progress. Yet it may
be declared that the personal influence upon these men is
dynamic and far-reaching. Some of them have doubtless
been able to regain that sense of harmonious relationship
to all existence which the non-human parts of life, unin-
fluenced by intelligence, possess. For, with the develop-

ment of intelligence in man, this sense of harmony was increasingly disrupted because man could adapt himself satisfactorily to his environment only by understanding, and his understanding was very meager. The development of knowledge has represented the increase in understanding. The era of specialization was necessary, at least in the investigating field, to bring this understanding almost as competitive nationalistic exploration after Columbus was probably necessary to bring understanding of the globe. With only parts of understanding no sense of harmony could be regained. But these men in the vanguard today are able, for the first time in history, to understand almost fully and completely the endless evolutionary forces which have brought them, their world, and their " universe " into being. They are able to some degree, to understand how and why they think; to understand not only the historical and sociological influences which have shaped them and their societies but all the vast and complicated evolutionary process of which they are the result. Intelligence has gained understanding expressed in generalizations, composed of words, symbols which, when used with scientific precision and definition permit the vital energies of man to expand as harmoniously as the vital energies of a bird soaring in the air or of a flower growing in the sun. Some of these men in the vanguard have such a sense of freedom and release as could never have been possessed by human beings since intelligence first evolved and man began to be filled with the fears, illusions, and superstitions which must oppress him until the understanding we have described could finally evolve. Note in this connection the following quotation from " Foundations of Psychiatry " by Dr. William A. White:

The false distinction between mind and body has gradually given away to a method of approach which no longer stresses this distinction but sees in the mind, the personality make-up, the final expression of the total integration of the individual into an organic unity . . . the conquering of the environment is made possible only by an increase in the knowledge of that environment which is, speaking in general, accomplished by a process of becoming conscious of the things which constitute the environment and of the laws that govern them . . . the human mind may profitably be considered as an instrument for contacting with the environment . . . the two concepts, individual and environment, far from being mutually exclusive, can only be considered as the two elements of a dynamic relationship of a constant inter-play of forces, in which their relative values are in a constant state of flux.

The integration of knowledge has resulted for these men in the vanguard in " an increase in the knowledge of that cnvironment . . . accomplished by a process of becoming conscious of the things which constitute the environment and of the laws which govern them," using the word " environment," of course, to express the totality of phenomena apprehensible to man. In the case of these men in the vanguard " the human mind . . . an instrument for contacting with the environment " has been used to obtain the harmonious adaptation to existence as a whole which men from the beginning have sought. The sense in which the words " the new scientific knowledge " are used will be made clear from all the foregoing attempts at definition. It will be seen why this new knowledge has been described in Chapter II as " the most potent and valuable possession of mankind."

If such an effect can be made upon the minds of the few men who have been able to grasp all the essentials and to realize the implications of the knowledge which this generation has acquired; if the largely unconscious applications of mere uncoordinated parts of this knowledge has resulted in "the conquering of the environment," it is easy to forsee what the eventual effect of the widest possible efficient diffusion of this knowledge to the race as a whole might have upon the happiness and well-being of men and what unpredicted "conquering of the environment" it will permit to future generations.

Part of the result of the application of uncoordinated portions of this knowledge has been the rapid development of the new devices of intercommunication which have brought a very large proportion of civilized men into such close and continuous touch with each other that the practically instantaneous, world-wide dissemination of facts and ideas is now possible. The whole modern world is more closely tied together than any small principality of the past. Today, information can be transmitted more quickly and more easily from New York City to Melbourne, Australia or Capetown, Africa than information, one hundred years ago, could be transmitted from the Battery in New York City to the sections a mile or two away.

Even the physical movement of men has been accelerated to such a degree that a circuit of the globe has recently been made in less than twenty-eight days. The minds of the great masses of the population are fed daily in a way to which nothing in the past history of the race is comparable. Through schools, churches, and similar older institutions and through the printed pages of newspapers and magazines; through telephone, telegraph and cable; through motion pictures and radio the minds of men in all civilized countries are kept continuously supplied with information and ideas.

To such extent as to be almost negligible has the realization of the potent new scientific knowledge of which we are now in possession been, as yet, transmitted through all this intricate machinery. All the illusions and imaginings of the past find daily universal expression. As parts of commerce the devices of intercommunication and the instrumentalities for the dissemination of facts and ideas are largely in the possession of private capitalists closely allied with the political and economic masters of the modern world. These instrumentalities are, accordingly, as yet used largely for the entertainment and amusement of the general population and to infiltrate into the minds of this population the information and ideas which possessing and ruling classes desire promulgated.

To develop the somewhat fanciful figure of which a hint was given in Chapter I when the few isolated scholars of the ancient world were described as bearing to mankind almost the same obscure and unperceived relationship which the human cerebral cortex bears to the human organism as a whole; we may intimate that the widespread, closely cohered body of scholars in the vanguard of contemporary progress comprise an evolving cerebral cortex for the organism of human society; that the devices of intercommunication represent the evolution of a central nervous system for that social organism and that increasing research and investigation represent its special senses. Primitive man had a cerebral cortex and a central nervous system but was actuated and motivated mostly by the subconscious instincts and passions necessary to his survival.

Conscious of the implicit fallacies in any such simile or analogy, we might intimate that the organism of human society today is in something of the position of primitive man. Consciousness of world-wide homogeneity is only dimly in being. Endangered nationalism and individualistic competition necessary for mere survival are still among many similar influences affecting the contemporary world. Just as the central nervous system of primitive man carried and transmitted impulses based upon fear and instinct, so these vast new devices of intercommunication still carry and transmit impulses based upon fear and instinct. However fanciful this figure—used only for its suggestive value—may be, there are other figures which symbolize the existing situation more exactly.

The progress of the race is always like the advance of a phalanx. There has always been a cutting edge comprising the men who were dissipating ignorance by research; who were in the fore-front of the race progress through the darkness of ignorance. The cutting edge of the phalanx today comprises the world-wide body of scientists coordinated through international congresses, books, theses and all the various devices which permit advanced scientific thought to be internationally known. Back of this cutting edge, of this vanguard, the phalanx widens out to include the many hundreds of millions of human beings living on earth contemporaneously with the men who comprise it. As has been stressed throughout these pages, these populations have been profoundly affected by the devices resulting from the advance of scientific knowledge; devices established and perfected by inventors, engineers, and capitalists and, to large measure, controlled by forces having no direct responsibility to the people nor under any effective social supervision by governments. It appears, almost, as if between the small groups in the vanguard and the great mass in the body of the phalanx a huge wall had been built up isolating realistic and dynamic learning from the mass of the population and blinding those in the vanguard to a sense of their great social responsibility. Back of this wall, facing the body of the phalanx, possessing centralized and coordinated control of the devices of intercommunication stand the great newspaper proprietors, the owners of the vast system of motion picture production and distribution, the monopolies controlling radio, even, to some extent, certain persons unquestionably representing an attempt to manipulate the general educational system in the interest of an economic oligarchy.

All the richest and most powerful civilized countries of today show that the actual control of the vast machinery for continuously reaching the minds and imaginations of the population is in the hands of extremely small groups of men who are very closely in touch with each other. In the United States of America, for example, nearly all book publishing, magazine publishing, and centralized news ser-

vice editorship are centered in New York City. This is
also the seat of centralized control of credit and capital.
It is the seat of centralized ecclesiastical coordination.
Most of the potent and dominant figures in all these activities touching the lives of one hundred and twenty million
people, yet centralized in New York City, are personally
acquainted with each other and have numerous occasions
to meet and to discuss their interrelations. Because of the
very nature of the activities with which they are concerned
these men are rarely likely to be of a type or temperament
appreciating or really respecting the importance and
potency of scientific knowledge in the realms of pure
research.[3] They represent new developments of the dominant new social forces which were depicted in Chapter II.
Occasionally when the learned world is stirred by great
new discoveries or theories the news is carried through the
machinery for disseminating information which this closely
cohered group controls. As in the case of all other news,
the attempt is made to stress the sensational and dramatic
aspects of these discoveries or theories. The general population gains an increased respect for the mysterious realm
of " science "; is momentarily made conscious of the
existence of the vanguard of the social phalanx but, having
no broad mental framework into which to fit the new facts
nor any rounded understanding which would permit true
comprehension of the new theories, no lasting social results can ensue. By the integrated groups dominating
political, economic, and social life in modern democracies
the vanguard of scientists is regarded largely as a group of
" impractical men " from whose strange and usually futile

investigations there may occasionally come something of enormous " practical " importance. Because of this likelihood of occasional extremely valuable contribution the whole group of scientists is treated in outer observance with profound respect, which occasionally masks a slightly contemptuous and patronizing attitude.

The individual scientist in the vanguard has been able to win to his understanding and eminence only by a lifetime of the most arduous and painstaking thoroughness. He has watched the pernicious social consequence frequently resulting from a few careless or hasty words on the part of some fellow scientist. He has come increasingly to understand the need for the most relentless precision of expression. He has come to realize ever more keenly the enormously complicated nature of all the phenomena investigated by science. He finds it increasingly difficult to make clear to untrained men the conclusions to which his patient, lifetime work have brought him. He finds himself perplexed and annoyed by the dangerous implicit fallacies and errors in the writings of " popularizers " seeking to give the general population some understanding of the work of himself and his confreres. For progress in his own speciality, unceasing concentration upon his researches is necessary. Inevitably he loses something of the common touch, he finds himself ever less interested in the reflection of contemporary life given by magazines and newspapers, he finds himself unable to give the attention necessary to understand the reality underlying contemporary political movements and bringing into prominence conspicuous political personalities. Naturally, he is drawn to those men who are equipped to understand his work—his fellow scientists throughout the world. Every day brings him reason to doubt the capacity of the masses of the population to understand the work with which he is concerned. When, occasionally, he seeks to lend assistance to movements apparently seeking social betterment he is generally perplexed or shocked by the obstacles, intrigues, and ignorances he encounters. However strong his sense of social responsibility, he finds it difficult or impossible to make any effective contributions save in his own work. He realizes that he is performing a highly specialized social function. He comes generally to feel that those concerned with the other highly specialized social functions of government, religion, journalism, education, are performing their tasks in the same spirit in which he attempts to perform his own and that any methods by which he might effectively cooperate with them are difficult to perfect and require thought and energy which his own activities do not allow him. As an actual problem confronting every scientist this situation and this conclusion seem inescapable. Yet, the fact remains that the body of thought with which these scientists, as a world-wide group, are concerned is relatively disinterested and comprises the most enduringly precious possession of mankind, while the activities and concerns of those dealing with government, finance, commerce, and all the other complicated general social activities of man are inevitably influenced by considerations arising from self-interest. They are, moreover, socially valuable only to the degree in which they are influenced by the broad social vision which can result only from the knowledge gained from these men in the vanguard, insulated from the great mass of the population and isolated from the daily concerns of men by the wall which all the considerations we have stressed have served to erect.

It would appear the task of statesmen, of the general academic hierarchy and, especially, of those in control of the great financial endowments charged with the advancement and diffusion of knowledge, to visualize this contemporary situation and to recognize the necessity for perfecting means of breaking down this wall between the vanguard and the rest of the phalax so that the great new instrumentalities of intercommunication which have recently evolved could consciously, articulately, and deliberately be used for the more efficient diffusion of scientific knowledge. In fundamental essence all these new devices are cultural artifacts representing potential new educational agencies. As such cultural artifacts statesmen should give them the closer scrutiny and attention which they deserve; should recognize their far-reaching and incalculable potentialities. By the centralized authority of such men as Hadrian and Trajan in past ages, having different outlook and different political nature, these cultural artifacts would certainly have been used consciously as instrumentalities of social betterment. In our democratic modern world no similar centralized authority with far-reaching power exists. Yet, our modern societies are in possession of social agencies which have been evolved in an attempt to retain the social vision which truly great statesmen of older social organizations possessed and to serve social functions of this type.

The new knowledge we possess and the new instrumentalities for diffusing it which we have evolved and are evolving, are social goods and social artifacts comparable to the crude knowledge and the crude social artifacts of which primitive tribes knew themselves to be in possession. It is the duty of the socially conscious portions of our modern world to see this fact clearly and to seek to utilize the great new things we know and have, at least as well as the chieftains of primitive tribes sought to use the things they knew and had. Science must evolve its own new type of statescraft. Scientists must fully realize their social responsibility and must perfect methods of rendering science more socially useful.

IV

THE GROWING REALIZATION OF THE
CHANGED SITUATION

There have developed in the past twenty or thirty years
many national and international organizations, the very
existence of which implies a growing realization of the
new situation resulting from the great and sudden changes
with which we have dealt. Only a specialist devoting
years to the subject could enumerate all the new agencies
of this type. We may here point out the existence of a few.

Under the ægis of the League of Nations is The International Institute of Intellectual Cooperation coordinating
the work of able groups in most of the member nations of
the League. The International Institute deals not only
with attempts at coordinating international scientific activities but studies carefully, and seeks to organize to some
degree, the functioning machinery of journalism, motion
pictures, and radio throughout the world. This shows a
conscious or pragmatic realization of the infinite potentialities in these devices which form the central nervous system
of modern civilization.

Especially to be noted is one of the foundations given
quarters by the International Institute of Intellectual Cooperation. It is " *Pour La Science* "—*Le centre internatio-
nale de synthese*, in Paris. The purpose of this organization is to further and coordinate scientific research and to
render it more fruitful by diminishing narrow specialization and the utilitarian aspect of science.

It is possible that these apparently subsidiary activities of the League of Nations may eventually prove as important and even more enduring than the political purposes of that organization. These subsidiary activities will almost certainly be continued whatever the future destiny of the League may be.

The Soviet Government has, for many years, issued a weekly " Bulletin of Cultural Relations with Foreign Countries" and the Nationalist Government of China issues a somewhat similar weekly bulletin.

Before the formation of the League of Nations, an elaborate and highly sensitive new organization of the activities of the international scientific world had been permitted by modern conditions. There are international associations in each of the great major sciences, and international conferences of these bodies are regularly held, the reports of which are printed and widely circulated. The books and theses of savants at all the world universities are distributed internationally and seen by the scholars of all civilized nations.

The British Association for the Advancement of Science and the American Association for the Advancement of Science serve to bring together not only all the outstanding scholars of Great Britain and the United States but also many thousands of laymen who are interested in science. Each of these organizations has one or more publications of wide circulation reaching many thousands of people. Each of these organizations has regular yearly conventions which are very thoroughly reported by the press of the entire world. Similar organizations exist in most civilized countries.

The press of the world keeps in close contact with scientific developments and gives immediate world-wide publicity to all striking new discoveries or theories. Dr. Banting's work in discovering a new treatment for diabetes was given the same publicity by the more intelligent sections of the American press that sensational divorces or crimes are daily given in the less intelligent portions. Every effort was made not only by newspapers and magazines but even by motion pictures to give popular widespread explanations of Einstein's theory of relativity. Great and continued interest was displayed by the press of the world in Howard Carter's sensational discovery in the Valley of the Kings of Egypt. Almost every week recently there have been accounts of new archaeological discoveries throughout the world. It is interesting to note that four or five extremely handsome books about the Khmer civilization of Cambodia and the great ruined cities remaining from that civilization were recently published in the United States. New types of books popularizing scientific discoveries are now being published and finding wide markets as well as being translated into many different languages.

In England are the officers of a World Association for Adult Education with prominent men of various nations on its Board of Directors. In the United States of America there have recently been organized several adult education associations seeking to coordinate the activities of the many bodies concerned with the great problem of adult education.[4] In each of the United States of America are "Educational Associations" while similar Associations exist in many other countries. International conferences are regularly held by " The World Association of Educational Associations " and many of the activities of this body appear forerunners of the general range of international activities, the necessity for which this book seeks to point out. Twelve " International Associations of Education " are noted by the American " Educational Directory." In the more technical scientific fields, especially those of medicine and hygiene, the Rockefeller Foundation functions internationally.

The Carnegie Corporation is entrusted with the legacies left by Mr. Andrew Carnegie and is charged with " the advancement and the diffusion of knowledge among the American people." The activities of more than nine thousand public libraries in the United States and Canada are somewhat coordinated by " The American Library Association " of which many of them are members. This central organization not only concerns itself with all matters of library practice but has recently, through the activities of its " Commission on Adult Education and the Library," undertaken far-reaching experiments in the adult educational field particularly by the distribution of reading courses prepared by specialists and distributed throughout the library system. There are over three million members of the General Federation of Women's Clubs in the United States of America. This organization has a committee charged with education and devotes much thought and effort to various activities seeking the diffusion of knowledge. The American Federation of Labor has considerably over two million members and its numerous publications deal frequently and ably with educational matters.

A large proportion of the young people of the United States of America are influenced by the activities of such organizations as The Young Men's Christian Association, The Young Women's Christian Association, The Boy Scouts, The Girl Scouts, The Camp-fire Girls, and many other national institutions of this sort. All of these organizations concern themselves in some degree with the questions of education and the diffusion of knowledge. There is in the United States of America a national organization known as the Congress of Parents and Teachers which has local branches throughout the country. This organization is designed to bring closer relations between parents and the teachers of the children of these parents. There have recently been instituted in America quite a number of "Worker's Colleges," organized and supported by various sections of organized labor.

It is claimed that in Denmark many of the political and economic difficulties which confront other modern peoples have been obviated by the far-reaching influence of the specific type of secondary school known as the "Danish High Schools." There have also evolved in Denmark the so-called "People's Colleges" which have attracted international attention. The success of the "People's Colleges" in Denmark has prompted experimentation along similar lines in the United States and several "People's Colleges," making a novel approach to adult instruction, have been initiated. Many of these organizations have been very successful and all of them are designed to recast curricula so that the working class will be given instruction believed to be of maximum value to it. An organization in the United States known as " The Worker's Education Bureau " seeks to coordinate all the numerous educational activities of labor papers, labor unions, worker's schools and other agencies being continuously evolved for the instruction of working men and women.

There is a Bureau of Education which forms part of the Department of The Interior of the United States and has a " Home Education Bureau " which prepares reading courses in many different fields and seeks to reach great numbers of people. The " university extension " activities, which are of very recent origin, have developed so rapidly that there is a " University Extension Association " in nearly every one of the United States and these Associations are coordinated by a National Association of University Extension Directors. The " summer schools " which now give higher instruction to many hundreds of thousands of people each year are also of relatively recent origin.

Listed in the " Educational Directory " published by the Bureau of Education, United States Department of the Interior, are twenty-five educational boards and foundations seeking various educational purposes. In addition there are a very great many church educational boards and societies.

Books on education flow steadily from the presses of all countries while more than 200 " educational periodicals " are published regularly in the United States of America alone. Within the past twenty-five years there has been the most striking success of so-called " correspondence schools " giving instruction entirely by mail. There is a national organization of commercial producers of "educational motion pictures" and there are numerous developing bodies studying the whole subject of "visual education" through the means of motion and still pictures. Novel publishing experiments have been made in recent years and have proved enormously successful. More than fifteen million copies of the "Everyman's Library" have been sold in Great Britain, Canada, and the United States while similar series of books issued by other publishers have had very great success. The new "book-clubs" in the United States distribute many hundreds of thousands of able books, some on scientific subjects, each month. The use of modern advertising methods and of the American periodicals of enormous circulation to sell books by mail-order must not be overlooked. One publisher using these methods claims to have sold within five years one hundred million booklets selling for a few cents. Many of these booklets not only contain the best of classic and standard literature but give a succinct digest of various aspects of scientific knowledge. In Tennessee the most efficient printing plant in the world has been established. The paper used is made in an adjoining factory from which it is fed directly into the printing presses. Books are here printed with a cheapness never before approached and are distributed in unheard of quantities through the vast chain stores and mail-order enterprises functioning nationally. As yet the type of book published is selected because of its existing widespread appeal as a household favorite, but this new mechanical efficiency will unquestionably figure in the general matter of future diffusion of knowledge.[5]

Plans are being made by one of the great radio broadcasting stations in America to conduct a " radio college " having lectures sent from a centrally located broadcasting station while classes to listen to these lectures will be gathered by county representatives all over the country and discussion stirred among the hearers by these representatives.

It is interesting to note that even the " comic strip " of the popular newspapers—a social evolution against which many educators have previously inveighed—is being used in an attempt to give children and others historical, literary, and scientific information.

Throughout the United States are numerous private schools giving primary and secondary instruction along novel experimental lines. Many of these schools are co-ordinated by an association known as the " Progressive Education Association " which has yearly conferences and publishes a magazine entitled " Progressive Education." The " New Education " movement in England follows sómewhat similar lines.

In California a movement has started which will doubtless spread very rapidly into many other states. A Director of Adult Education has been secured to coordinate the activities of all the existing agencies dealing in any way with attempts at adult education.

Among the numerous new journalistic developments of recent years have been the many publications popularizing science, giving information of all new inventions and discoveries and appealing specifically to the desire for scientific and technical information. It is not without interest, in this connection, to note that the "National Geographic Magazine," the monthly organ of The National Geographic Society, has a circulation of over seven hundred thousand.

The day before he went on the operating table, in May 1921, Franklin K. Lane wrote a letter describing his vision of a "super-university"—"a place of exchange for the new ideas that the world evolves each year." He wrote:

No faculty—but a super-university with all the searchers and researchers, inventors, experimenters, thinkers of the world for faculty. No students—but every man the world round interested in the theme under discussion welcome as a student without pay.

Such a super-university in some form or other will doubtless develop in the future. It will diffuse advancing knowledge throughout the world from one central international group of great minds in the vanguard of human progress.

The educational potentialities of older institutions such as museums of natural history, fine arts, and industrial arts are being constantly broadened by the work of such bodies as The American Association of Museums. Many of the American Colleges and universities are now establishing machinery for "alumni instruction." President Hopkins of Dartmouth said in his inaugural address so long ago as 1916:

The College has no less an opportunity to be of service to its men in their old age than in their youth if only it can establish the procedure by which it can periodically throughout their lives give them opportunity to replenish their intellectual reserves.

Nearly all the agencies and activities which have been noted are of relatively new development. Nothing quite like them was known in the past. They indicate that all classes of the populations of modern democracies realize the importance of knowledge and seek to develop instrumentalities for helping in the diffusion of knowledge. For the first ten years after the World War it was everywhere obvious that the thought embodied in H. G. Wells' much-quoted phrase, " It is a race between education and catastrophe," was profoundly influencing world activities.

As a whole the new agencies and activities show an ever-increasing recognition of the potentialities of the new devices of intercommunication and of the desire and capacity of all portions of the population to gain broader understanding than is now possessed by them.

The following significant quotation is made from the introduction to a series of books entitled " Chemistry in Industry " published by the Chemical Foundation of America:

The ease with which any of the natural sciences can demonstrate how indispensable it is to modern life is simply a testimonial to the complexity of our civilization. However important the various sciences and the fields of engineering which deal with their application may be, it is certain that chemistry and mathematics are the really fundamental positions.

This quotation shows a constantly developing recognition by so-called " practical " men of the value of pure science and pure research. A very wealthy American has recently endowed a botanical experimental station where the research workers are freed from all pressure of producing " practical results " and are left free to devote a lifetime, if necessary, to research work having no apparent practical value. President Hoover, when Secretary of Commerce, was instrumental in aiding efforts seeking to secure a large fund of many millions of dollars to be used for the endowment of workers in pure science. Slowly the fact that disinterested scientific endeavor is absolutely essential to the progress of the modern world is being recognized even by the successors of the " practical " men who were formerly so contemptuous of integrated knowledge and of research and so insistent that educational institutions give " practical," specialized instruction.

The minds of the present generation have been continuously assaulted with various political, economic and social programs, doctrines, and panaceas. There can be no doubt that there is a constantly increasing number of intelligent people in every country who come ever more to distrust all doctrines and programs and to believe that the only enduring welfare of the race is dependent upon the more efficient instruction of the great masses of the population.

We have seen numerous well thought-out idealistic endeavors given, by chance, an opportunity to test their validity. And we have seen that, regardless of the intrinsic merit or demerit of these programs, they have largely failed because the great masses of people, whose clear understanding of them was essential to enduring success, had no mental framework permitting realistic understanding of the aims sought or the plans for achieving those ends. Lester Ward long ago pointed out that the only hope for enduring social progress lies in the better understanding of the people and that, in consequence, the task of the diffusion of knowledge is the foremost problem and responsibility confronting our time.

As yet the results of the many new instrumentalities which have here been dealt with are relatively meager and the growing recognition of the value of pure science is limited to an almost negligibly small proportion of the political and commercial worlds. Nevertheless these new developments show an obvious trend. The results will be more marked when the many activities which at present overlap each other, or are conducted along too narrow lines, are better integrated and given a clear sense of purpose and direction.

Already, pleas showing a recognition for this necessity for coordination and for formulation of purpose and plan are forthcoming. There have been suggestions for a "Senate of Humanity" to be composed of outstanding scientists of all lands integrating knowledge and seeking to bear to the modern world something of that relationship which the tribal chieftain bore to primitive tribes, in the sense of keeping clear understanding of the true nature and value of the knowledge of the tribe as a whole. It is not without significance in this general connection that the great masses of China and Japan evince a marked desire for motion pictures, automobiles, radio sets, cameras, fountain pens, and similar productions of modern science. In these countries moreover, in many instances, the most ultra-modern devices of transportation and communication are being installed. China carries much of her mail by airplane and it has been predicted that China will probably jump from the ox-cart stage to the airplane and motor truck stage without going very greatly through the stage of railroad transportation. The influences of these devices, resulting from Occidental science on the Oriental mind will be far-reaching and profound.

Education has had no basic revolution. Our contemporary educational systems are largely an inheritance from the past. New accretions have been made to them but no great national or world-wide attempt has been made to revolutionize them completely and to substitute for the largely unplanned and inefficient existing system a new system planned as political constitutions and industrial structures are planned; a new system fitted to the modern world and the needs of modern people.

The time is obviously approaching when fuller and more dynamic realization of the new situation will develop, the existence of new conditions, requirements, and potentialities be recognized and world-wide efforts be made to apply to the study of the question of the diffusion of the new knowledge, the scientific method of which that knowledge is the product.

V

THE NEED FOR BROAD VIEW, SCIENTIFIC
METHOD, AND DEFINITE PLAN

The impact of the new knowledge upon the minds of intelligent men and women throughout the world has been very great. The recognition of the potentialities in the new devices of intercommunication has been profound and far-reaching. Both of these have resulted in endless sporadic and not definitely correlated efforts to use the devices of intercommunication for the diffusion of parts of the new scientific knowledge. They have resulted in myriad empirical changes and developments in educational material and technique and in practically unceasing criticism of details of the existing educational system. We are beginning to observe, moreover, the efforts of many intelligent and progressive men to utilize newspapers, motion pictures, and radio for broadly conceived educational purposes or experiments. Only to almost negligible extent has there evolved, however, any recognition of the necessity for basic criticism of the existing inherited educational system as a whole. And it cannot be contended that there is general realization of the completeness of the entirely new situation confronting the race as a whole; the entirely new situation brought by the great and sudden changes with which we have already dealt.

The impact of the new knowledge, the realization of the new potentialities have not yet brought a broad view of the new situation nor realization of the need for a basic new philosophy of education which must underlie and motivate any effective new machinery evolved to deal successfully with the new conditions, requirements, and potentialities. A broad view on an international basis is requisite since the situation is one which affects all mankind and not any particular part of mankind. The world is, now, obviously one. The knowledge we have is the final crystallization of the efforts of men of all races and ages for hundreds of thousands of years. It is in the possession of men at points everywhere throughout the world. The new devices are utilized throughout the world. The influence of the proper diffusion of the new knowledge will affect the entire race for long generations to come.

No enduringly valuable results can be hoped for from quick, so-called "practical" activities. The problem is really the transforming of the combined folkways of all the contemporary world with its myriad peoples of radically different environment, traditions, and background. The results sought cannot be hoped for save after generations or centuries. Moreover, the task is not really so much the promulgation of new facts and ideas for themselves alone as it is to seek the disruption of inhibiting illusions, ignorance, totems, and tabus which can only be enduringly disrupted by the facts and ideas gained by scientific investigation.

The broad question of the diffusion of knowledge cannot be approached in the spirit in which all political, economic, and commercial problems are usually approached.

The only analogous undertaking is represented by such attempts as were made by the American, French, and Russian Constitutions seeking to shape and influence the activities of many successive generations. The task is one which offers no place for personal ambition, unrationalized humanitarian or idealistic emotion, no personal or group theories or doctrines, no selfish purposes of any kind. The task is an objective one calling for all the qualities of great statesmanship, of wide vision, and of disinterestedness which the race possesses. The task is one requiring the cooperation of the ablest and best informed men in every department of human activity and in every one of the great divisions of learning. In the past the folkways of individual peoples, the unrationalized customs and accumulated illusions, prejudices and habits, were disrupted most frequently by the thoughts of one man or a small group of men effective only after many centuries. It has been declared, for example, that Socrates disrupted the folkways of Athens, that Christianity disrupted the folkways of the Pagan world, and that the Renaissance and numerous contemporary activities disrupted the folkways of Medievalism.

We are now growing conscious that all the inherited empirical and pragmatic knowledge of the entire race; all the infinitely diverse inherited customs, habits, and prejudices of all the peoples of the earth must, eventually, be disrupted; that a future very greatly shaped by man's conscious aspiration, intelligence, courage, and will must be brought into being. We are now growing conscious that, in the evolutionary process, there have evolved in man faculties and capacities which permit him, to increasing degree, to shape himself and his environment somewhat in accordance with his own will and that his success must always be partly dependent upon the clear-cut purpose and wide understanding which he brings to the task. The Romans of the period about 500 B. C. were only very dimly conscious of the results which would eventually ensue from the utilization, during eight hundred years, of the unusual energies and capacities which Rome was beginning to exercise. Even Paul himself could have been only dimly conscious of the results which the promulgation of Christianity, during many ensuing centuries, would bring. Francis Bacon could certainly not have dreamed of the results of the ideas he wrote into the *Novum Organum*. Milton, Locke, Voltaire, Rousseau, Diderot, Jefferson, and Tom Paine, were only dimly conscious of the far-reaching political and social changes which would result from the widespread promulgation of the ideas they advanced. But we, of today, seeing the rapid changes which have resulted from the application of random bits of modern scientific knowledge; seeing the results upon the outlook of the relatively few men who have grasped the implications of the new scientific knowledge of which we have come into possession, are able more completely to visualize the results which may ensue during thousands of years to come from the perfecting of broad, far-visioned, and far-reaching plans for the most effective scientific diffusion of this knowledge to the largest possible proportion of the race. The inevitable result will be to bring a new world into being. And it is for that reason that scientific method is requisite so that the situation may be envisaged objectively, methodically, and entirely free from any nationalistic, sectarian, or doctrinal preconceptions. Only a plan based upon the fullest possible understanding of the relationship of the new scientific knowledge, as a whole, to mankind and all the existing institutions of mankind as a whole, can hope for maximum possible success.

So stated, the task seems visionary and hopeless. Yet, to our generation it actually involves no more than attempts at formulation of a basic new philosophy of education developing from full realization of the changed conditions and of past human experience; the preparation by an international commission of recognized international scientists of a minimum curriculum comprising the basic essentials of scientific knowledge considered as an organic whole; and the consequent, conscious, and articulate efforts of all organizations, concerned with the diffusion of knowledge, to make the formulation of this new philosophy of education and this minimum curriculum known to all the highly centralized and closely coordinated forces which now form actual or potential parts of world-wide machinery for the general diffusion of science. There are great parts of the task which we must leave to the future but we can, at least, prepare the way for farther reaching future activities based upon knowledge still to come from research.

Until today we have thought of education too much in terms of instruction; too much in terms of formal guidance given from the fifth or sixth year on to some varying period of " graduation." We begin, increasingly, to see that education is a life-long process, that it begins with birth and ends only with death. We begin to see that the impressions absorbed by a child within its first year or so are of the profoundest possible importance. Not only do the folkways surround every human being as fully and continuously as water surrounds the fish which live in it, but we begin to see that the psychic structure of newborn children is as much a product of endless evolutionary ages as is the tangible, anatomical, and physiological structure— the complicated evolutionary genealogy of which we come ever more to comprehend and to respect. It will be many generations or centuries before our knowledge has become sufficiently comprehensive or our energies sufficiently coordinated to deal effectively with these basic aspects of life-long education.

It will be many generations or centuries before the more effective diffusion of realistic knowledge so influences the folkways that the complex cultural artifacts affecting children from the moment of birth will be so changed as to alter such early impressions of the external environment as now serves so greatly to shape the general psychic life, the tendencies, motivations, emotions, and mentation of the people of today. But the recognition that we must leave to the future these and many other great parts of the task should help us to clarify the portions of the undertaking with which we are already capable of dealing. In broad essence this might be considered as the attempt at construction of a simple new brain pattern to replace the many diverse brain patterns now affecting the minds of the race. Thus expressed, of course, endless implicit fallacies inhere in the statement. An attempt will be made to show recognition of these implicit fallacies and to develop what is meant by " brain patterns."

One of the primary characteristics of scientific knowledge is that it represents such coordination of experience as to cause increasing doubt of the primary impressions received by the senses of the individual. The individual's eyes give him much reason to believe that the world in which he lives is flat, that the sun unquestionably goes around the world. Full realization that the earth is not flat and that the relationship of the earth to the sun is very much more intricate and involved than appears can result only from the integration of ideas gained by the accumulated experience of thousands of men over many centuries. It involves, moreover, the construction in the individual imagination of some more or less valid picture of the general position of the planet in the solar system and of the solar system in space. The construction of this picture cannot result from the information gained by the individual's senses alone; it can only result from the facts and ideas gained by the individual from study of the records giving the crystallized, disciplined, corrected, and tested experience of mankind as a whole.

Similarly there is a very obvious difference between a man living and the same man dead. It has appeared to men from the very beginning that something left the dead body. This something is immediately considered as a mysterious entity worthy of consideration and is given the name of the " spirit " or the " soul." From the skies thunderbolts are apparently hurled and the simple thought processes of primitive man visualized another something out in space which must have hurled the thunderbolts. After long ages of magic and pantheism the idea of monotheism evolves and identification is made between the mysterious, intangible, and imponderable " soul " or " spirit " and the mysterious, irascible, and capricious being assumed to hurl the thunderbolts—an omnipotent divinity which must be placated or praised by elaborate religious procedure.

To the individual man there is a marked difference between that part of his being which he uses in chopping down trees, which he must support with food and water, and that part of his being which he uses in expressing words and communicating ideas and thoughts. Soon there is the general acceptance of a division between " body," " mind," and " soul." Abstractions have been objectified. And on the basis of these objectified abstractions a brain pattern vaguely explaining the mystery of existence develops in the individual mind. This general brain pattern affects to some degree all the men accepting the many diverse religions and philosophies existing today. Science, refusing to objectify abstractions, unconsciously attempting to distinguish clearly between the reality which can be apprehended by human " senses " or abstracting capacity and ideas bearing no verifiable, communicable, and demonstrable relationship to reality, cannot, of course, accept this type of brain pattern. By slow and patient development of all the numerous sciences it has come, after tens of thousands of years, to see that man is an evolved organism resulting from the most infinitely complicated

evolutionary processes over a hundred million years; that
the earth on which he lives is a relatively small planet
moving by natural laws through a vast solar system which
moves, in turn, by great natural laws through interstellar
space the immensity of which is ever more clearly made
apparent.

Science has to refuse to see man as a simple mixture of
" body," " mind," and " soul " but has, instead, to regard
him as a highly complicated organism as a whole, of which
thought is but one of the forms of behavior. Science has
to refuse to concern itself with such objectified abstractions
as the " soul " and with such assumed absolutes as the
omnipotent divinities of which this " soul " is assumed to
be part.

Permitting the assumption of no absolutes; crystallizing
out of all knowledge, of all human experience, only the
generalizations which formulate understanding of natural
laws, formulations capable of demonstration, verification,
and communication, science has to represent a relatively
very small area constantly expanding outward in the sense
that it acquires more detailed information and constantly
expanding inward, not toward any absolute of any kind
but toward increasing, coordinated understanding of the
relationship of phenomena and of the laws influencing
phenomena. In 1930 this small area of relatively exact,
verifiable, demonstrable, and communicable understanding
may be declared to constitute a small area of relative certainty the general outline and basic essentials of which can
form a brain pattern giving to those to whom understanding of it comes, a new mental picture of themselves in

relationship to the general range of mystery in which they find themselves. The progress of the race has, in a sense, represented the constant change of the mental pictures of this relationship in the heads of men.

For many centuries large proportions of the people of the Dravidian races have had a mental picture of the universe in which the world on which they found themselves rested upon the back of a huge turtle floating in a sea of milk. The religious books of all the older races represent attempts at explanation of the mystery of man's being and his relations to the phenomena he sees about him and the past from which he has, in some mysterious way, resulted. No savage tribe has been found without some more or less crude cosmogony. Most parents find that children, almost from the first acquisition of speech, ask the so-called ultimate questions.

In one sense all the philosophies and religions of the world have been an attempt to allay the questionings of the race concerning ultimate questions by affording explanations in which, as we have pointed out, there was always an assumed central absolute and, moreover, an intricate mélange of objectified abstractions resulting from subjectivism. Science relentlessly seeks the answers to the ultimate questions which every child asks, which every savage tribe has asked, which all philosophies and religions have sought to explain, which the race has always pondered over.

The essential difference between science and religion and the older philosophies is that it seeks to purge itself of subjectivism, refuses to assume absolutes, or to objectify abstractions. It methodically, and with a minimum of preconceptions, investigates phenomena. It records its data and from these data evolves generalizations which are immediately subjected to criticism and experimentation. These generalizations which prove immune to criticism, which stand the test of experimentation, are accepted as being—at any given time—relatively and approximately correct explanations of the natural laws affecting phenomena. Science seeks to arrive at understanding that which is true for all men of all kinds at all times everywhere; it deals only with phenomena which can be made apprehensible by ever developing devices permitting an augmentation of the special senses. Science seeks always to see everything in relation to everything else and refuses to deal with generalizations which do not take into consideration this complicated relationship.

Certainly the difference between a so-called intelligent and educated man in the daily intercourse of contemporary life and of a so-called ignorant man is that the intelligent man has a greater realization of the complexity of everything that he sees and touches and hears about; has a crude mental picture based on certain aspects of the crystallized knowledge, the evolution and development of which during the past few thousand years was traced in foregoing chapters; while the ignorant man lacks this realization of complexity and has a mental picture resulting from his individual uncorrected sense experiences and from the implicit cosmogony gained from the religion or philosophy which he accepts. To the degree in which random bits of developing scientific knowledge come to these men from newspapers, magazines, lectures, motion pictures, radio, or other sources, it will be seen that the more intelligent and better educated man will have a better mental framework into which to fit the new facts or ideas, to understand their importance, to remember them and thus make them part of his mental equipment. While to the ignorant man these facts and ideas will seem part of the general chaos of ideas in which he finds himself floundering; and, having no mental framework into which properly to place them, they will be neither understood, appreciated, nor remembered.

Education proceeds from birth to death. Experience is gained from birth to death. But education and experience are effective largely only to the degree in which there is some mental framework into which all that is learned or experienced can be fitted. It is for that reason that use of any part of the actual or evolving educational structure to disseminate random facts in the various fields of knowledge cannot be really effective until minds are fitted for such integrated understanding of the basic essentials of scientific knowledge—the nature of man, the nature of his processes of thought, the nature of society, of the earth, of the general position of the earth in space and time (objectified abstractions the implicit fallacies in which are recognized)—that new facts fit properly into place and thus acquire and maintain value as mental possessions dynamically influencing the person gaining them.

The attempt to furnish new brain patterns based upon modern scientific understanding unquestionably influences the educational systems of all contemporary civilized peoples. But these systems are vast and intricate. They comprise tens of thousands of teachers of all types, kinds, and classes. There are no central coordinating agencies consciously and continuously directing the work of the educational systems as a whole to this end. And it is partly for this reason that the end is obviously not achieved. It is obviously not achieved for, were it achieved, the civilized countries of today could not possibly present many of the peculiar aberrations which their political life and their social institutions daily display.

There is a central motivating purpose to the religions which shape India and to the Confucian philosophy which shapes China. Vast and intricate agencies and instrumentalities have been evolved over long centuries to promulgate these central motivating purposes. But from the distance of thousands of miles we can see that the purposes become confused and that the great masses of the populations of India and China translate them into crude superstitions.

We assume that the scientific understanding which constitutes the power and individuality of the Occidental civilizations is successfully infiltrated through all our vast and intricate devices for reaching the general population. But, for lack of clear formulation and of continuous effective coordination, this assumption is obviously fallacious. Students go from kindergarten through colleges and even, occasionally, through postgraduate schools and emerge with only unintegrated, confused, and befuddled understanding. The portions of the new and dynamic knowledge which have been given them have been permitted to

be hopelessly intertwined with illusions, ignorance, preconceptions, and dogmas more analogous to the superstitions of the masses in India and China than we like to
admit.

The mental picture possible to a man properly instructed
in modern science cannot be combined with the mental
pictures resulting from the philosophical and religious cosmogonies remaining as vestigial relics in our modern world.
The incompatibility is too great. Unless the mental picture
resulting from science is clear-cut and integrated it is
relatively valueless. If it is confused with the dogmas and
superstitions of the past it results in personal unhappiness
and social misery. The problem of the scientific diffusion
of knowledge, the great task now confronting the world,
is to free instruction from the vestigial remains of the past.
This can be accomplished only by the clear-cut formulation
of a clearly seen purpose and by the evolution of world-
wide coordinating bodies articulating a basic new philosophy of education and completely replanning contemporary
curricula. The desired brain pattern, the desired mental
framework, is in the heads of the men we have described
as being in the vanguard of contemporary progress. It has
recently been subjected to the new integration resulting
from the electronic theory and other developments. It has
been made more dynamic and freed from some implicit
absolutisms by Einstein's theory of relativity. But none of
these affect it so radically as to negate any of those essentials of understanding of phenomena which comprise it.

Each day more men throughout the earth undergo intellectual experiences which help to develop a new, relatively

homogeneous, brain pattern in their heads and to purge away beliefs, illusions, and dogmas remaining from the past. The result upon the daily life of the men to whom this understanding, based on science, comes is great. It affects their relationship to those close to them, to their children, to their associates, and to the world at large. The fact need not be stressed that children born to highly educated men with sound scientific understanding are less influenced through all the seconds, minutes, hours, weeks, months, and years of infancy and early childhood by the intricate folkways which influence children born into the families of men without such understanding. With the increase of the number of men who can be given or who are able to obtain this understanding, the pressure of the existing folkways will find ever stronger obstacles and barriers.

This point may appear minor but is made in order to indicate that our present lack of understanding of the congenital psychic structure; our realization that purely cerebral understanding is but one of the aspects of human mentation and behavior need not deter us from attempting more efficiently to seek to disrupt the diverse, prevalent cosmogonies; to widen the range of influence of contemporary scientific understanding and to influence society to whatever degree may be possible by more definite and deliberate attempts to have existing and evolving educational systems give the type of brain pattern inevitably resulting from integrated understanding of the new knowledge.

Scientific method necessitates that there be as precise

as possible definitions of purposes and clear realization
of more or less arbitrary limitations.   The limitations may
be arbitrarily set as including within them only the range
of facts and ideas which no recognized scientist would
undertake to dispute.   The "universes of discourse" of
scientists vary greatly.   The detailed differences between
scientists are very great.   But no scientist of today will
deny that the world is approximately round and is certainly not flat.   No scientist of today will deny that man
is an intricate organism, the product of all past evolutionary processes.   No scientist will deny that our planet
is part of a relatively small solar system moving by relatively well understood laws through interstellar space, the
vastness of which is recognized.   No scientist can fail to
see that the evolution of human society and civilization
has been very slow.   These are typical of basic essentials
which are hardly likely to be made matter of dispute.
However great the differences between scientists they are
all at least certain that most of the beliefs contempor
aneously affecting the vast proportion of mankind are not
true.   And the range of understanding which permits them
to be sure that these beliefs are not true is the very small
area of relative certainty which this book desires to stress;
it is the central core of integrated scientific understanding,
the necessity for the more efficient and widespread diffusion of which this book seeks to point out.   Here, then,
we have approximated a description of the "scientific
knowledge" which an international commission should
integrate and use as the basis for a minimum curriculum.
    Scientific method must also be applied to the study of

the nature and potentialities of all the instrumentalities through which this knowledge could be diffused.

Without question the modern world has endless devices for instilling facts and ideas into the minds of all civilized men on earth almost from the moment of birth to the moment of death; for connecting the individual mind—the cell of the social organism—as fully and continuously with thought, the blood stream of society, as the individual cells of the body are connected by the actual blood stream of the body. These devices, separately or as a whole, are not consciously and articulately working to afford all minds everywhere brain patterns into which all that is learned throughout life can be fitted. Scientific method should study them in great detail separately and as a whole; should study the possibilities of so coordinating their activities that they should work, as far as can be made in any way possible, as one great race instrumentality.

Once a basic new philosophy of education is formulated, once a minimum curriculum comprising the basic range of facts which should assuredly be given to every child born on earth is made available, the whole educational structure can function more efficiently and all parts of the existing and evolving educational structure can be clearly seen in their true relationship to each other.

From the new viewpoint which would be rendered possible the kindergarten, the primary and secondary schools could, for example, be designed as parts of a world-wide, life-long educational system giving, at least, such brain patterns that the desire for knowledge would be awakened, the libraries would be used and newspapers, motion pictures, and radio of a new type developed to give sound, detailed information continuously to the greater number of people who would appreciate the value and understand the true nature and place of this information. The present rather mechanical instruction could be given a more organic quality. Great modern buildings and ships are designed in this general fashion, the details being evolved to fit into a definite plan which has been conceived. The elaborate political and social organization of the United States of America is, in a sense, the outgrowth of the more or less definite plan which found expression in the Constitution of the United States—a plan which has regulated the activities of five generations.

All successful human activities require a clearly formulated purpose and a definite plan. It is by such purpose and plan that gold, iron, coal, and other physical things are obtained and distributed throughout the world. It is by such purpose and plan that all the commercial activities of the modern world take place. Knowledge is, of course, intangible. The institutions seeking to diffuse it deal with the least obvious and most intimate relationships of people over many long years of their lives. The very thought of a definite plan for the diffusion of knowledge seems like the effort to plan something hopelessly and eternally intangible and complicated. But all the educational institutions which have ever existed have been motivated by a purpose only vaguely formulated.

This book does not seek to formulate any dangerously simple and easy program or plan. It does seek to throw—from many different angles—light on the general nature of the contemporary situation and to intimate that, however intangible knowledge may be, the attempt at least can be made to seek its scientific diffusion with something of that broad view, scientific method, and definite plan which all other human activities require for success. The problem may well be considered as one in social engineering. The attempt to solve it today is made by what is largely empirical procedure. Some day the problem must be approached as architects or engineers approach the solution of problems presented them.[6] There is no reason why the first efforts along this avenue should not be initiated by this generation.

VI

THE POSSIBLE AIMS OF THE EDUCATION
OF THE FUTURE

Since pansophism for the individual is now obviously impossible of realization, and since the experience of over a century has proved that specialized instruction without broad general understanding does not produce a completely satisfactory type of individual or of society, a new philosophy of education appears requisite. Since the word " philosophy " has fallen on evil days we will, however, speak rather of the " aims " of education. It is detail which has augmented. The ultimate questions confronting mankind are the same as always and are very simple. They touch every human being closely and every human being asks them at some time during life. From the philosophical point of view (we use the suspect word in a historical sense) all progress in science and knowledge represents—in the last analysis—only the obtaining of more satisfactory answers to these ultimate questions through ever more intricate and complicated researches and understanding.

Since, then, for individual happiness and wholesome, continuous, and satisfactory growth full realization of the relationship to individual well-being of knowledge as a whole is requisite, while for social well-being highly specialized knowledge and capacities are necessary, it appears that the future may have only to break down the existing undeserved reverence for shallow technical train-

ing and to give the individual student—before specialization begins—a broad view of the ultimate questions
asked and the answers to them gained by modern science.
The effect of this upon the individual mind is to give a
new sense of confidence, a fuller sense of personal freedom
and personal responsibility, and enduring release from
those perplexities and fears which mysticism and dogmas
have, up to this period of history, sought to assuage. Even
specialized technical knowledge is rendered more dynamic
when clearly seen in its proper relation to the human
adventure as a whole.

Each individual—very early in life—can be made to
realize that there is no absolute, fixed knowledge; that
knowledge is ever-augmenting and that no man or group
of men has a monopoly on it; that every individual may
very properly be looked upon as a representative of the
entire race, can stand at the center of the knowledge which
the whole race has gained and see that from that center
broad avenues radiate out to the ever advancing frontiers
of knowledge.

Under the largely empirical procedure of today instruction begins with the disciplines which the past thought
necessary preliminaries to any acquisition of knowledge.
The individual child is lead slowly out from study of its
immediate surroundings to study of its own country, then
to some history of the past. Coincidentally, information
and instruction totally uncoordinated and utterly arbitrary
and mysterious are given to the child. There appears no
attempt to give the child any understanding of what the
aim and purpose of the whole long years of instruction is

to be. It is rather as if a man who was to be taken for a trip over the entire world were to be started on the trip without being supplied with maps and charts; without ever being given understanding of the general nature of the world with which he was to be made familiar. It is assumed by contemporary primary and secondary educational procedure that integrated understanding will inevitably result from the acquisition of detailed information, even if a broad and high mental framework it not afforded. We have little evidence that this assumption is correct. Unintegrated detail causes intellectual chaos. Without a clear plan, the individual dashes madly hither and yon throughout his life. Scrambled brains result from scrambled facts and ideas.

Scientific procedure would appear to require that the youngest child be regarded as an intelligent being and that it be quickly made aware of the aims and purposes sought by the instruction which will be given it throughout life. It is, after all, being started on a journey of exploration and discovery. And this primary and basic fact should be made clear to it so that its personal interest and cooperation may thereby be more effectively elicited. Everything taught it should fit into a rudimentary, but ever enlarging, mental framework. The difference of the effect of such instruction and the effect of the present day instruction which, without explanation of their nature or usefulness, forces certain facts, certain capacities upon the children, can be instantly envisaged.

The present day system causes the reaction and the withdrawal of the child from what appears to it something incomprehensible and arbitrary. It is designed as if the child were a mechanism rather than an organism. The center of individual interest, zest, and enthusiasm is not reached and stimulated. Rather, the natural desire to grow which the mind possesses in common with seeds and all living things is adversely affected. Vital intellectual energy is stifled, not freed. If we accept Dr. White's dictum that " the mind may profitably be considered a device for contacting with the environment " it becomes obvious that the vital energies of the child in relation to its environment cannot be released except along avenues of thought and understanding. Contact with the environment is made by human beings through understanding developed by the use of symbols—language. As John B. Watson says: " Man is the only animal that takes the universe to bed with him." Meaning, of course, that the words and images—symbols—in man's memory and imagination form a symbolized " universe " within him.

A child's failure to understand any personal value or importance to itself in what is told it inevitably serves to insulate it from that particular range of information. This is of value if it protects from trivialities or the really unimportant. But, by current systems of education, whole basic fields of thought are thus made to appear remote and forbidding to the child. The wider and subtler ranges of information developing from some basic field of thought or fact are thus, during the ensuing period of maturity, also insulated from the child's understanding. This means that the harmonious expenditure of the child's energy and the harmonious development of its interest and understanding are prevented.

For Rome to hold power and to grow and use its vital energy, roads had to extend in every direction from Rome. For any individual mind to grow and use its vital energy, roads of understanding to the remotest frontiers of " the area of relative certainty" must be established in that mind.

Today knowledge is, from the beginning, made to appear arbitrary and remote from life. Few men in the modern world have escaped this confusion of ideas, because prevalent types of instruction resulted in a sense of insulation of " life " from " knowledge." Knowledge— instead of being a word representing to them the rich soil in which the mind can grow and attain fulfilment and happiness, a means of self-expansion and self-realization— is, from earliest childhood, made to appear something remote from the undefined word " life " or the ideas of " growth," " happiness," and " joyousness." The fact that so many American colleges have recently initiated " orientation courses " is one proof of this statement. The academic hierarchy has obviously begun to realize that the uncoordinated, detailed instruction given students by many different courses leaves little enduring effect since it appears too remote from the personal interests of the individual. The " orientation courses " seek to correct this condition by giving the individual a new sense of his relationship to contemporary civilization, to all life and knowledge and thus to render more dynamic the specialized information he may receive. It would appear, however, that this attempt at orientation might better be made at the very beginning of instruction in early child-

hood than during the course of higher instruction to
which only a relatively small proportion of the population attain. At present nearly all primary and secondary
instruction represents arbitrary attempts to mold the
child mind by intellectual disciplines evolved by the past,
the underlying ideas and purposes of which are very
far removed from the new understanding of the nature of
children which modern research has given us. To most
teachers, moreover, the original theories on which the use
of these disciplines are based, are unknown. Consider, for
example, in this connection the modern teaching of arith-
metic in relation to the original theory upon which mathematics was taught. At best current higher education can
do little more than seek to undo some of the repressing
influences of primary and secondary instruction.

Another evidence of the developing widespread popular
recognition of the necessity for integration is the recent
remarkable success of many books which have sought
to afford such integration. H. G. Wells' "Outline of
History" was treated by the press of the world as a social
event of great importance. Numerous journalists declared
that from it could be gained a better understanding of
much of what man now knows than from long years of
instruction at American colleges. This book unquestionably served to integrate for many thousands of people the
scattered historical information of which they were in possession but the close relationship of which to themselves
had never before been made so clearly obvious. J. A.
Thomson's "Outline of Science," while it lacked the
chronological interest and the integration of Mr. Wells'

volume, was also received with acclaim and very widely circulated. We have recently had "outlines" of philosophy, art, literature, and of all human activities in quantities so large and of merit so varying that the press has very justly begun to burlesque and satirize the attempt to reduce knowledge to a few volumes or even—in one extreme case—to a single volume! The minds of large sections of the reading public have, however, been beneficially influenced and stimulated by the best of outlines and there has even been recently the development of a new type of volume represented by Dr. George A. Dorsey's "Why We Behave Like Human Beings" and Dr. Paul de Kruif's "Microbe Hunters" seeking to give integrations of much more detailed and highly abstract portions of scientific understanding. The fact that there is a wide interest in and a wide market for these books is the best possible proof of the increasing recognition by ever larger numbers of people of the need for integrated understanding. But it is obvious that only a very special scientist or popularizer will have the unusual combination of capacities permitting the preparation of a book which, while being sound scientifically, will also make wide popular appeal. What should be given by wise life-long study cannot be given by a few books, no matter how excellent.[7] At least the success of these numerous attempts at integration and the world-wide interest in them shows that even the general public realizes that something is wrong and indicates that there would be a very marked international response to the efforts of an international commission which would attempt to outline the most important things which the race now knows and to prepare a minimum curriculum which would not only serve to direct the work of popularizers but would also orient general educational activities.

A properly designed minimum curriculum and the new educational material and technique which would evolve to instill it would very probably help in a few generations to bridge the gulf between "knowledge," and "life," and "happiness," which now inhibits and prevents the growth of so many millions of minds. For, inevitably, in integrating knowledge it would inexorably humanize knowledge as the various outlines have sought to do. With the whole educational structure—from the very beginning of instruction—seeking to make clear that knowledge is only the tested and disciplined accumulation of all that men of all ages and times have been able to learn concerning their nature, their origin, the nature and origin of the planet on which they live, the nature of the universe in which that planet moves, the means and methods of making human life more secure, more satisfactory, happier and more rhythmic and harmonious; it is certain that the contemporary common feeling that knowledge is something apart from life could not survive. Rather even the youngest child could be brought to feel that instruction was designed to give it something infinitely precious, and with better formulations of the nature and humanist value of knowledge even those in charge of primary instruction should be able to give to the youngest pupils mental brain patterns which would prove acceptable and stimulating. Even the introduction to instruction would view knowledge as an organic whole. All instruction would seek to release rather than stifle the organic forces of curiosity, zest, and enthusiasm.

The underlying philosophy of the education of the future may probably accept the basic biological generalization of the adaptation of life to its environment, regard human intelligence as the device for making satisfactory adaptation to environment, envisage each individual human mind as potentially part of the race mind, and recognize the necessity that every individual mind should be early afforded a mental framework based upon the essentials of knowledge and permitting steady, continuous, and harmonious mental development throughout life since detailed understanding given by life and study would fit clearly, usefully, and dynamically into this framework. Without such framework, filled in as individual growth proceeds, the ever developing devices of racial intercommunication can be used only to appeal to instinct and passion, not used to diffuse knowledge, correct aberrations, and tie the minds of men into an effective race mind.

Long ago, even before the development of modern biology, political theorists had begun to use the terms "organs of government" and to refer to the "organic law." Throughout contemporary political writing is a growing tendency to refer to "social organisms." The organic nature of societies is being recognized and there is evident inclination to begin to envisage society as a continuing whole, as an organism. In this conception the individual must be regarded as a mobile, intelligent cell of the organism of the race. We have already gone so far as to intimate that modern science is one of the forms of behavior of what may fancifully be regarded as the evolving cerebral cortex of the race organism and we have intimated that the evolving devices of world-intercommunication are a central nervous system for the race. We have gone so far as to refer to the interchange of thought throughout the world as the circulation of the blood stream of human society. These figures are obviously fanciful and will, unquestionably, be offensive to many scientists because there are sinister implicit fallacies in all similes of this type. Yet analogies are useful and the general underlying thought in a possible philosophy of education of the future may be some recognition of a modicum of correctness in these similes prompting the formulation of ideas seeking so to influence educational activities and procedure as to bring each child, each man and woman—the mobile intelligent cells of the race organism—consciously within the circulation of this social blood stream of thought. The protoplasmic cell is equipped to absorb nourishment from the blood stream of the organism of which it is part. The chief effort of the education of the future may be to equip the individual human being to absorb mental nourishment from the great rapidly flowing stream of thought.

The possible philosophy of the education of the future will probably seek to permit the most harmonious possible adaptation of the individual organism as a whole to human environment as a whole—in the sense of the totality of apprehensible phenomena—by permitting the individual energy to expand harmoniously out through that organic body of knowledge as a whole which comprises the generalizations obtained by the human mind " a device for contacting with environment " in all relatively successful attempts (at all times, everywhere) to make satisfactory " contact." This body of knowledge, of course, is what we call " science." Any prophecy of such a philosophy of education must necessarily be defended by contemporary work in science in regard to the nature of man and the nature of knowledge.

Elliot Smith has synthesized the work of many biologists showing that man is an evolved organism fitted to gain ever fuller and more effective awareness of environment. The recent advances in many natural sciences permit Korzybski's definition of man as an abstracting organism crystallizing ideas—abstractions of various orders, from experience—a definition, of course, made in relation to this Polish mathematician's basic definition of man as a " timebinding class of life." The devices and capabilities comprising his abstracting capacity give man the ability to " bind time." [8] The basic generalizations of the Gestalt psychology also appear to support the prophecy made. Modern psychiatry stresses the inexorable close relation of the individual to all human life and achievement past, present, and future. Knowledge is the accumulation of all that men have learned. The various disciplines of the sciences and of education have evolved to permit intelligence effectively to abstract from this race accumulation as the senses abstract from the various phases of the physical environment.

It is becoming possible to define " man " and " knowledge " in a way never before conceivable, in relatively objective and scientific terms. Such definitions are requisite for the proper formulation of a sound philosophy of education.

In a certain sense it may be declared that the evolution of life marks the evolution of awareness of environment, that the evolution of the sensory apparatus represents a detailed awareness of environment; that the evolution of the cerebral cortex represents the evolution of a conscious awareness of environment and that the modern scientific knowledge which we have sought throughout this book to show as a slow crystallization represents a *dynamic* awareness of environment. It is this dynamic awareness of environment which education should seek to give. The philosophy of the education of the future will seek to render individual and race intelligence ever more dynamic. It may quite possibly combine the impossible ideal of pansophism and the obviously insufficient and defeatist theory of specialization by teaching—to everyone—the basic essentials of scientific knowledge as a whole before giving special instruction in any one field of useful activity. It must almost certainly set as the goal of primary instruction among the people of all races the perfecting of a minimum curriculum giving to every child born on earth what may be described as an anatomical chart of the organic body of science so that the facts and ideas ever more vigorously and continuously afforded by evolving instrumentalities of intercommunication can be understood and, properly placed; used to educate—not disintegrate.

Education—once organic but limited—totters because bulk now exceeds framework. Scientists—by scientific method—must build a new mental skeletal framework for the education of the future.

Man is an energy system. Education is a device for affording outlets for that energy in ways beneficial to the individual and to society as a whole.

VII

THE POSSIBLE MATERIAL OF THE EDUCATION OF THE FUTURE

One of the most effective ways of arriving at a reasonably satisfactory solution of the problem of the educational material which should be diffused through the existing educational system and the new devices of intercommunication is to seek to envisage—from an ideal viewpoint—the possible material of the education of the future. By " material " is meant the facts and ideas abstracted or crystallized from all race experience which will be selected as the range of facts and ideas of maximum value, interest, and usefulness to the individual and, hence, to social harmony and well-being. It may be somewhat confidently predicted that this material will comprise a humanist integration of scientific and cultural knowledge from which all irrelevant and unimportant details have been cut away. Let us consider a few basic generalizations which may consciously underlie the educational material of the future if it is deliberately prepared by an international commission undertaking a problem in social engineering; or which may consciously underlie this material if it is empirically evolved.

Whatever else the individual human being may be, he is an organic center of energy. Like all organic things he seeks expansion and growth. The growth and expansion of individual human beings is, however, rigorously limited in what has been called the "physical realm" although practically illimitable in what has been called the "mental realm." When the distinction between mind and body is no longer accepted in terms of gross empiricism and the general sensory and cerebral activities of man are regarded only as parts of his general "abstracting" capacity, this whole question of growth and expansion is seen from a new viewpoint.

The harmonious and continuous growth and expansion of the human being is dependent upon understanding. Growth represents the expansion of the individual energies out into the environment to which, in consequence, satisfactory adaptation is made. This growth cannot take place without understanding. Where there is no understanding growth ceases. From this point of view it will be seen that passionate acts are energy expressed in physical violence and destructive emotionalism rather than in wholesome growth bringing satisfaction. Endless criminal activities of individuals, and indeed of nations, result from human energies inevitably seeking outlet and failing to contact satisfactorily with the environment through understanding. They are attempts vainly to crash through by the use of energies on levels not characteristically human but shared with many forms of life lacking the essential human characteristics. The relationship of propaganda here will also be very obvious. But note, moreover, that when the inexorable attempt is made by each human mind to understand and so to grow on the characteristically human plane, growth is inhibited not only by "ignorance" but also by the accumulated folkways; the complicated interplay of totems, tabus, and dogmas resulting from organized absolutisms inherited from the ignorance of the past.

The attempt is here being made to picture the individual human organism almost from birth as a center of energy hemmed in almost completely. Conscious of the fallacy of objectifying this idea of energy, let us nevertheless, momentarily, attempt to keep a picture of something which seeks to escape from endless complicated toils in which it has been taught. There are iron bars and granite which utterly impede this energy. There are apparently soft strings and threads which appear to give, but give only very little. From the beginnings of human intelligence the lives of most men who have lived on earth have represented this unceasing fruitless attempt to break through all these impediments. And the great names in history are the names of those men whose audacious and disciplined intelligence served to break through or to cut through these impediments and to carry great masses of their fellow human beings with them. Whenever a human being or a human society is, or has been, static, it is because the growth of that being or that society is, or has been, prevented by accumulated factors preventing understanding, thereby inhibiting growth and thereby serving to affect the individual or the society as the "breaking of a horse" affects the horse or as the planting of a seed in a pot too small to permit its full growth affects the plant growing from the seed.

In what is frankly an attempt at prophecy where the imagination is thrown ahead several centuries, we may be permitted to deal with human beings, not to be born for long generations to come, from an unusually objective point of view. From this point of view we must view man as only a class of life; view man in much the same terms that we would view fish or birds or animals. As a " time-binding class of life " possessing practically all of the capacities of all other forms of life yet having certain unique characteristics, all the capacities of man, including thought and speech, may be viewed as modes of behavior remarkedly similar, as contemporary biology is continuously stressing, to the behavior of so-called " non-intelligent " forms of life.

Highly complicated artifacts can be produced by man without " thought " consciously formulated in words. Though, consciously expressed in words, is necessary for more intricately evolved artifacts, and, especially for the general complex of elaborate political and social institutions affecting the lives of civilized peoples.

A child born in a jungle in some part of Africa, South America, or New Guinea not affected by civilization will grow to maturity adapted, to a large degree, harmoniously to its local environment. It will, however, be inhibited and affected from the moment of birth by systems of tabus and inhibitions more complicated than any civilized man well appreciates. All these tabus and inhibitions will be associated with ceremonials and rituals in which *words* will be used in the attempt to formulate the accumulated experience of the tribe and the accumulated efforts at the tribe's understanding of the general mysteries encompassing it. Harmony is immediately lost if external forces change the environment or if the native is moved to some dissimilar environment.

A child born in a city in one of the highly civilized countries of today will follow a course very much similar to that of the young savage save that its chances for physical well-being and for the escaping of quite tragic experiences will be less. This civilization in which the second child develops is to a very great degree merely the accumulation of ever more intricate artifacts but its institutions have grown enormously complicated, detailed, and far-reaching. All of its institutions of government are based on the *words* of its constitution and its endless array of statutes. Its ecclesiastical institutions are based upon *words* in books alleged to be of divine origin. The more complicated devices which increasingly influence the daily lives of civilized peoples are based upon scientific generalizations expressed in *words* written in the books of researchers and scientists. Those who would be masters of these institutions must thoroughly understand them. The *raison d'etre*—frequently lost to view or not made clear—of all educational institutions is to give this understanding of the nature of civilized institutions. All those desiring " success " for themselves or their progeny seek to gain or to transmit this understanding.

Every great capital observes the rise to power, economic, political, and social, of men adapting themselves to this highly complicated environment—not the result of " nature " but the result of the accumulated time-binding capacities of humanity—by mere vigor, alertness, and flexibility of mind but without clear-cut and articulate understanding of the real nature or the remote background of the complicated institutions which form civilization.

And the eventual collapse of these men is almost inevitable. For modern civilization represents a great tangle, a great jungle through which one can move unscathed only by clear-cut understanding of the origin and development of all the various intricate and interacting institutions which man has built. And, beyond this comprehension, must be the comprehension of the general evolutionary process in which man and his civilization are such a very minor and temporary manifestation.

The energies of the human individuals, then, can be expanded harmoniously only by understanding. And understanding can come only through *words*. Where there is a word which is wrong all history shows that vast intricate institutions evolve to create new inhibiting factors to human growth. For in every word there is an implied system of metaphysics. Every word is an inevitable outgrowth of the complex of experiences and thought which produce that word. One cannot talk comprehensively in the year 1930 without using such words as " relativity," " tropism," " psychiatry," " electrons "—a whole vocabulary evolved in the past few years to deal with new discoveries resulting from that dynamic awareness which is science. Progress always comes bearing a new vocabulary—the new ideas need new words in which to find expression. There is no phase of ultra-modern thought more interesting or more profoundly significant than the augmenting attempts of mathematicians to afford means of giving words precise definition so that they can be exact symbols permitting agreement and rigorous demonstration. The work of Alfred Korsybski, C. K. Ogden, W. N. Whitehead, C. J.

Keyser and others in this field is attracting wide attention. Words have become the walls of the prison of many modern minds. If not used with scientific precision they develop into static doctrines which confine intellectual energy as completely as dungeons confine prisoners. The instant that realization of this fact is appreciated, and understanding that most words are " variables " is gained, doctrines can become " doctrinal functions "—the *relative* and *variable* nature of which is recognized.

Every word used in religion and metaphysics contains inherent and implicit *absolutist* explanation of the mysteries of existence. To understand the sphericity of the earth the disruption of any absolute " up " and " down " was necessary. To understand Einstein the disruption of any absolute " time " and " space " is necessary. To escape from any dogma understanding of the invalidity of the absolutes inherent in that dogma is necessary.

It is the value of well defined words used precisely by scientists that they deal only with processes and relations and, in consequence, cannot—to those who properly understand them—become prison bars for intellectual energy. The education of the future will assuredly appreciate more greatly than the past has done the profound importance of the words used in giving instruction.

So, attempting to visualize these children to be born several centuries from now we see each one of them as an enormously vital organism possessing practically illimitable energies and illimitable capacities for growth and development. We see that the type of awareness which they share with other forms of life cannot possibly give them the requisite understanding which will permit the proper expansion and outlet of the energies they will possess. We see that for such outlet—harmonious and continuous—the deliberate development in them of " dynamic awareness " resulting from the crystallization of disciplined race experience in generalizations in which *words* are used with scientific precision, will be requisite.

" Culture "—in whatever sense the word is used—represents the satisfactory expenditure of complicated human energies. In its finest manifestations—the arts, literature, and music—it represents the attempt to adorn life, to make life more gracious and beautiful by changing the conditions affecting life in accordance with some deep-rooted human effort at rhythm and completeness. Whether of the individual or society it is an attempt at harmonious growth, the realization of which is possible only through understanding. Consider phases of some of the cultures of the past. The almost incredibly perfect—and entirely useless—bits of handicraft procurable in India are the result of the idea that the making of one perfect thing during life would cause the " soul " of the maker to be reincarnated on some higher plane because of the satisfaction rendered Buddha by this perfection. In other words, human energies of the most highly developed kind were expended on utterly useless artifacts because of an idea—" understanding " of some fallacious theory of the universe. Consider the great ruined city of Angkor. Here is a vast city built concentrically out from a small sanctuary in which the god, Siva, was assumed to reside. The city had probably five hundred thousand people. It is estimated, on the evidence afforded by the ruined structures, that the priests and their attendants who were devoted to the worship of Siva totalled at least seventy-five thousand. The conflict between the temporal power of the monarch and the ecclesiastical power of the priests of Siva is evidenced in all the buildings—the ever increasing magnificence of the palaces and the ever increasing malignity of representations of Siva designed to heighten the influence of fear of this non-existent divinity upon the monarch. Here is only one of the countless examples which all the past affords of the energies of whole vast peoples over centuries being given to the development of a culture affording all the excellent things of life to monarchs and priests but leaving the people in wretchedness and squalor. Any European or American visiting the Orient is struck by the contrast between the palaces, the temples, the miles of intricate carvings, and the " degradation " of the masses. Where kingship and divinity were identified as, for example, in Egypt and Imperial Rome, much of the culture represented merely attempts to placate or gratify this center of authority and its favorites.

Back of the economic oppression of European feudalism which influenced, of course, the whole culture of the Feudal era was the idea of the " divine right of kings." The assumed absolute, God, through the intermediation of the Pope had given the king or feudal lord dominion over the people. Inevitably the daily lives and the energies of the people were directed or shaped by this idea.

It need scarcely be pointed out that the culture of nearly all the past was profoundly affected by the absence of the idea of " progress " which is of recent origin, while all the culture of today (and note that it expresses itself more in transportation systems, great canals and bridges, huge office buildings rather than in the fine arts) is obviously the result of the prevalence of this idea of progress. It is not without bearing on this general question that we see the amazingly rapid development in America of endless devices designed, not for the gratification of rulers, but to make life easy and more pleasant for the individual. " Yankee inventiveness " was displayed almost from the earliest days. For the idea of equality and of individual liberty was accentuated by pioneer conditions and men found themselves free to use their ingenuity, imagination, and intelligence to devise better ways of harnessing horses, of making tools and implements, in developing new artifacts of daily, homely usefulness.

In all the instances which have been given, and in tens of thousands that might be given, it is obvious that the culture of any place or period or epoch has been the result of the understanding, of the general range of ideas influencing the place or period or epoch. Contemporary America shows, for example, a situation where the greatest motivating incentive is the desire for " success " which, in turn, is almost associated or identified with the idea of " wealth." The whole national life is influenced by this prevalent doctrine—the fallacies of which are best known to those who have been " successful " in this way only to find—to their perplexity and amazement—that this divinity " success " was as non-existent as Siva if happiness and harmony were sought as the result of its worship.

The contemporary development of preventive medicine, the unprecedented degree to which in recent generations, useless activities have diminished and the energies of man have been used to make human life more secure and satisfactory are indications that the education of the future will certainly be infiltrated with the idea of conscious adaptation or control of "progress"; and will seek to direct human energy into channels beneficial to the race as a whole. For the successful consummation of this ever-developing purpose, understanding is essential and the material of education of the future must be designed, or must evolve, to furnish this understanding. We can glimpse some of its probable aspects by viewing in very simple, humanist terms the books representing new integration which were mentioned in the foregoing chapters.

When Gibbon wrote his famous "Decline and Fall of the Roman Empire," history did not extend much beyond the beginnings of that empire, even to the minds of very learned men. In a sense men's minds, men's understanding, men's imagination and intellectual energy stopped a few thousand years back in time. Nothing could mark the basic changes during the past century and a half more than the fact that H. G. Wells' "Outline of History" regards Rome as but one of the relatively minor aspects of history and forces men's minds, men's understanding, men's intellectual and imaginative energies back tens of millions of years before Rome. So in the field of astronomy, the 1eld of the indefinitely large, every school child of today realizes that the planet on which we live is a very unimportant astronomical body, that our own sun is thousands of times larger, and that a star such as Alpha Orionis is almost incredibly larger. The understanding, the intellectual and imaginative energies of men and women of today are carried millions, hundreds of millions of miles out into space.

So, in the realm of the indefinitely small, Dr. de Kruif in his "Microbe Hunters" dramatizes this world of microbes and humanizes the knowledge gained by the men who have spent lifetimes peering into it by microscopes. Every school child of today knows that a drop of water is likely to be filled with millions of living creatures utterly unsuspected, undreamed-of by the understanding of men of earlier times. The same drop of water is, by the electronic theory, made clear to our contemporary understanding as a universe of energy and motion.

Note how science and research have broken down iron bars, blasted away the granite, cut through the cords and threads and permitted our understanding to range vastly out into a "universe" undreamed-of heretofore. The understanding—thus liberated—renders us immune to the fears of the man in the jungle. It permits a more dynamic release of all our energies. If properly integrated it lessens or destroys the hold upon us of current hysterias, current passions, current propaganda. It serves to sublimate energies which might, otherwise, readily find outlet in domestic dissensions, emotional spasms, neuroses, psychoses, and other maladjustments inhibiting well-being and happiness as well as growth.

Note how the importance of the educational material with which previous generations dealt is increasingly diminished and thrown into new perspective by this understanding. Alexander, Caesar, and Napoleon, cannot possibly appear so important to a well-informed man of today as they appear to men of even a hundred years ago. The cause of diseases which only recently appeared almost as afflictions of a malignant divinity are clear to us. The "heaven" and "hell" of earlier generations are not located by our developing geographical or astronomical researchers.

Yet note, on the other hand, that the inhibiting strength of prejudice based upon ecclesiastical explanations of the universe prevent the sane dealing with such pressing social problems as veneral diseases. It is obvious to any intellegent man that modern science—if unobstructed—could very rapidly diminish or even entirely remove these diseases which bring such untold misery in their train. Yet here is an instance where understanding is directly prevented from activity by organized ignorance based upon speculative explanations of the mystery of existence. A "god"—an assumed absolute has been assumed to inflict these diseases as a "punishment" for "sin." So long as persons holding such beliefs are able to influence opinion and legislation, objective understanding of the phenomenon as a problem in bacteriology and hygiene—no different from any other such problem—cannot be given to the general population. The dead hand of the past reaches out to prevent the human movement away from unnecessary pain and misery. It functions through doctrines—absolutist and erroneous explanations of phenomena—expressed in words harder to blast away than granite. It is a striking instance but there are tens of thousands of other instances in which the vital energies of the contemporary world cannot be released into channels which would diminish human wretchedness and increase human well-being because educational systems have not given—possibly have not been permitted to give—the understanding upon which these cultural developments could be based.

All the foregoing has been presented merely in an attempt to throw light from various angles on the possible nature of the educational material of the future. This material will probably be designed to give rounded and integrated understanding permitting the rounded expenditures of energies and the obtainment of rounded and rhythmic individual well-being. The direct relationship to the individual student of all the essentials of knowledge will be made clear by every device human ingenuity can perfect. Out of the experiences of the past will probably be crystallized the salient and vitally essential facts and ideas. Reverence for the past *per se* will almost unquestionably pass. Only the most vital elements representing continuous and unbroken developments and growths will be abstracted. Today so-called " facts " are forced into students' heads. In the future facts will certainly be considered only as contributing to understanding.

The basic ultimate questions man has always asked and the ever more satisfactory answers to them the race has gained and gains, together with the more effective knowledge which developed down the ages—not from consideration of ultimate questions, but from satisfactory and

effective adaptation to environment due to the ever augmenting understanding of natural laws and principles—
these will constitute the material of education.

Exploration of the earth's surface resulted from commerce, war, and many other immediate concerns of individuals or groups. The voyage of Columbus was partly
prompted by correlated knowledge gained by science but
much more vigorously motivated by the desire to reach
the gold and spices of the Orient. Exploration for its own
sake and for the acquisition of new geographical knowledge is a development of very recent times, Captain Cook's
voyages being among the first well-equipped undertakings
of this sort. The most striking instances, of course, were
the efforts to reach the North and South Poles. Over long
centuries the various discoveries resulting from diverse
interested motives finally brought the geographical understanding which we, of today, possess. And, today, that
understanding can be viewed with the complete disinterestedness of pure science and thought and seen as an
answer to one of the "ultimate" questions over which
men pondered. This is an analogy applying to practically
all the knowledge which we possess. The knowledge has
been slowly acquired in the effort to secure immediate
ends advantageous to those acquiring it. But the scientist
and educator—at any given time—can look upon the
complete accumulation. The motives actuating those whose
activities developed geographical knowledge are only of
minor importance to men and women of today born into a
world which possesses that knowledge.

From speculation based upon astronomical knowledge

science gave men of the fifteenth century reason to believe the earth round in shape. Science gave the brain pattern of this round earth divided into degrees of latitude and longitude. Science gave, also, the compass, and weapons superior to any possessed by non-Europeans. Actuated by every conceivable personal passion, instinct, desire, illusion, and idea, Europeans—after Columbus—sailed over all the oceans and moved over all the lands. Ultimate questions, the disinterested desire to augment knowledge probably actuated no single one of them. Nevertheless the result of their movement was to fill in the blanks afforded on the imaginary sphere by the imaginary lines of latitude and longitude. Man acquired detailed understanding of the earth on which he lives. The child, today, given maps of that earth need not—save for his personal pleasure and stimulation—be told much about the tens of thousands of boats which sailed; the passions, brutalities, shipwrecks, battles, or murders which attended these explorations. The essential result was the new answer to an old ultimate question—What is the nature of this earth on which man finds himself?

The same thought applies to all other understanding we possess. The largely unconscious application of characteristically human capacities has developed understanding without any consideration of " ultimate " questions. Yet, science and education can crystallize from these accumulations ever more satisfactory generalizations, answers to the age-old questions which all men have asked and all men will ask. To the degree in which science and education clearly recognize this opportunity they can function and develop most effectively. The knowledge is primary. The complicated details of how or when we acquired the knowledge is important but secondary.

The material of education of the future will probably comprise all the range of information which can best give the individual the most satisfactory and enduring understanding of a relative universe and of the place of this planet in such a universe; the origin and nature of the planet; the origin and nature of life, the development of life; the origin and nature of man's intelligence, the development of man; the origin and nature of society and its institutions, the development of society; the origin, nature, and development of knowledge—all leading down to the individual who can, by having gained the essentials of this range of information, begin to see and understand himself in proper perspective and to understand his real place in the great scheme of things. Every vital fact giving understanding of the actual functioning of the human body and intelligence would, of course, be part of the material of education.

The center of individual energy is the cluster of ideas which regulate and affect, influence and decide every activity of the individual. When these ideas are irrefutable and comprehensive, when they give full and complete understanding, the individual can adapt himself as harmoniously, as buoyantly and joyously to life as non-intelligent life is adapted by tropism and instinct. When these ideas are erroneous and limited the individual cannot adapt himself properly or harmoniously and unnecessary ultimate pain or disaster or futility inevitably impends.

When the brain pattern is a relatively true and objective mental picture of the individual and his human and extra-human relations, harmonious adaptation is possible. But it is not possible until all objectified abstractions and absolutes are completely blasted away and when there is an abiding and continuous sense of the enormously dynamic nature of everything which exists. One of the most pernicious aspects of contemporary educational practice is the failure to give this abiding sense of the dynamic nature of existence with the consequent hemming in of the vital energies of human beings by the wrong understanding that there is anything static in all existence. Coincident with the instilling of this range of facts and ideas indicated would be the instilling of knowledge concerning the greatest enduring achievements in all the arts so that the individual would be afforded the greatest possible opportunity to fill his life with the best that has been thought, made, or imagined by man. It is likely that such a future will develop arts unimaginable by us since they will represent that " exuberance " which William Blake declared to be art; an exuberance which can result only from the buoyant, joyous, and tumultuous outpouring of human energies no longer hemmed in at any point. In general it appears that the possible material of the education of the future will be that range of facts and ideas which can most and best inform, stimulate, stir, and free the individual and, while making his life happy, harmonious, and dynamic, make him realize to the fullest his great race inheritance from the past, his great responsibility to the future.

Granting any vital part of the ideas here advanced it

would appear that—in planning material for widespread diffusion through the existing, evolving, and expected agencies of education and communication—the suggested international commission would ruthlessly cut and prune away the dead wood which now obscures educational vision and devise material so vital, so stirring, so obviously valuable that the existing hiatus between the idea " life " and the idea " knowledge " would quickly and inexorably be bridged and men everywhere be brought to realize that the facts and ideas being given them were the only possible means of growing happily and harmoniously. Such facts and ideas would prepare the way for life-long continuity of study and bring realization that education must last from birth to death and that there is no " graduation " possible. A mental framework would be provided analogous to the hypothetical partly blank globe with its purely imaginary lines of latitude and longitude which existed in the heads of the learned men of the fifteenth and sixteenth centuries. Adult life with its experiences and its eagerly and zestfully sought information would fill in these spaces as adventurous experience filled in the blanks and gave us our modern maps.

VIII

THE POSSIBLE TECHNIQUE OF THE EDUCATION OF THE FUTURE

Since the future will almost unquestionably look upon education as a life-long process—the means of effectively and harmoniously merging the individual with the race—it will probably design or evolve a technique of formal instruction based upon the efforts to afford the individual maximum opportunity for effective and harmonious adaptation to all the multitudinous aspects of life and the universe—family, society, knowledge, the past, and the future.

Conditions of human life, up to this time, have necessitated that education be looked upon as a preparation for life. Henceforth—and ever more thoroughly until the civilized world will realize that education is, perforce, a life-long process—the idea of all formal education will probably be to get the individual mind started right and so awakened, stirred, and stimulated that it can and will expand indefinitely, harmoniously, and happily out into the world of knowledge as a whole. It will be recognized above all that the individual mind must be stirred and freed, that the individual must fully and clearly understand the evolution of the race as a whole, so that he can feel close identity with all intelligent life and come to regard himself as a mobile, intelligent, and dynamic cell in the evolving organism of mankind.

119

The technique of instruction will probably be designed for educational mechanisms so integrated, so interlocked, that they permit or bring the continuous expansion or unfolding of individual intelligence. In behaviorist terms all education will probably be regarded as a " conditioning " of the individual to fit into a society which is being deliberately and consciously shaped to regain that harmony which all the non-intelligent parts of life appear to possess, which man has temporarily lost because his social institutions were neither properly rationalized nor based upon clear understanding of the real and objective nature of man and society which the world is just beginning to acquire. Libraries, motion pictures, newspapers, radio, and similar devices can, in the future, be made the most important machinery of education, continually giving facts and ideas to fill in the mental framework given by formal instruction in youth.[9]

We can deal here only with the possible technique of elementary instruction the profound importance of which will, however, be clearly recognized. We have previously pointed out how the change in the folkways will probably increase the educational influence of environment on children long before any age at which they now begin kindergarten instruction.

Obviously " education " will not be regarded by the future as it is regarded today. By whatever means necessary or possible the distinction between " school " or " knowledge " and " life " will be completely broken down. Every effort will be made to have the child look upon the place of its instruction as that associated with its happiest moments. All our inherited ideas concerning the need for discipline will probably be revised when proper application is made of new educational devices which will not force the developing young mind to deal with arbitrary and dogmatic mysteries but will rather fill in these minds with the most exciting and interesting facts given through the medium of motion pictures and numerous other similar new agencies.[10] The mind of the child will almost certainly be regarded somewhat as a seed, as an embryonic, evolving, growing thing which must derive nourishment pleasurably from the rich soil of accumulated race experience properly plowed, harrowed, and fertilized by the tools of education.[11]

As the child's body has undergone a concise summary of organic evolution before birth, so after birth the child's mind will be brought through a concise summary of social and intellectual evolution with the result that—at maturity—the man or woman will fit dynamically and happily into society and civilization at whatever stage they have reached and will be fitted to help deliberately and consciously to advance society to higher stages.

Already it is possible to imagine very young children—their minds never having been subjected to blind creeds and dogmas—being given a new type of instruction from which all disciplinary and arbitrary ideas have been completely excised, from which all thought of reward or punishment has been removed; a technique designed to catch, awaken, and hold the vital interests and curiosities of the child.

The geographical knowledge of the race has been communicated through maps and globes. The anatomical knowledge of the race has been communicated through charts and models. In the great museums of natural history like the one in New York, involved laws like Mendel's law of heredity are made clear through actual specimens charted and diagramed. All the forms of life are shown either actually or, in the case of protozoa, in glass models.

It is interesting to observe quite young school children studying the highly magnified glass models of the circle of events ensuing from the biting by an *anopheles* mosquito of a person with malaria. The technique which is now being adapted by museums of this type to give comprehension of involved natural facts or laws to visitors is probably a forecast of the technique which will be evolved by the primary instruction of the future to make these facts and ideas comprehensible to very young children. The plans for the Temple of Science at the Chicago Century of Progress in 1933 include devices for showing the most involved discoveries of modern science.

Imagine children starting school where hanging from the ceiling would be a huge magnetized globe. These children would not be taught to read. Nothing would be *taught* them. Early instruction (the word is used for lack of the proper word which must later be evolved) would be entirely oral and by motion and still picture. It would be designed throughout to delight, please, and interest the child and would take into consideration all the psychic factors influencing the reactions of the child. As water and fertilizer are put on seeds and seedlings so the chief effort would be—not necessarily to touch the child's " mind " (again a better word is requisite)—but to fill the evolving memory and imagination with material upon which the mind, as it evolves and develops, can feed. Small magnetized figures would be placed upon the globe to give graphically the idea of gravity. And, doubtless, rather elaborate astronomical models would develop, graphically displaying the relationship of the globe to its solar system. It is easier to explain by "showing" than by "telling." From the beginning the idea would be firmly implanted in the child that life and knowledge are symbolized by the figures on the huge globe. The child—over a course of years—would watch the changes in the globe as all that science can learn of the past was shown by detachable metal plates or by motion pictures—depicting the growth of vegetation, the glacial ages, etc. Motion pictures would also seek to show the origin and formation of the globe. The origin of life would be shown through motion pictures carefully made under the constant supervision of world-wide authorities. When the story had reached dramatic forms of life—and proper intelligence can make the whole evolution dramatic to the child mind—small dinosaurs and other striking forms of early life would be shown on the metallic globe. Later the whole story of primitive man, the Paleolithic and Neolithic ages, the origin of every invention and institution—a connected, integrated, and harmonious story would be graphically shown. The contemporary first centers of civilization would be indicated, the slow spread of civilization, the slow bringing of the entire planet within human understanding and control. This process would, of course, take a number of years. The material used could be, to some degree, standardized and used in all countries. There would be continuous attempt at giving an integrated summary permitting the child to relive—as its body had relived from the moment of conception—the salient aspects of evolving life on this planet. The material would not be explained in any great detail—if at all—but would be implanted in the child's imagination and memory for it to draw on throughout life. Early impressions enormously influence us and are probably better retained in the memory than those of later years.

Coincidentally with the attempt at integration, would be attempts at analysis and explanation from the immediate moment of the immediate environment. Starting from everything vitally touching the child's life, man's productions for example, could be traced back. A teacher might break a drinking glass. "What shall we do?" she asks. "Let's make one." And the whole evolution of the making of the glass would be explained by motion pictures taking the child's mind back from the glass it uses to the time when glass was first accidentally made by the primitive man. So, with every object, every force or influence touching the child's life.[12]

From the beginning, part of the instruction would be the actual manipulation of increasingly complex mechanisms through such methods and models as those developed by the *Deutsches Museum* in Munich. And, from the beginning of instruction, the child would be given constant opportunity to see how all the essential work of the world is carried on. In the whole school life of the author, for example, there was never any contact with this tremendously stimulating range of fact. From six years to eighteen none of his instruction comprised actual contact with actual men and women doing the world's work, making the world's food, clothing, structures, and devices. It seems almost certain that attempts to make clear the development of man's control of or adaptation to his environment and the actual working of the machinery of that control or adaptation will be basic in the education of the future. The actuality has been dramatic to the race. The child may profitably be considered as a symbol of the race. The actuality can be made dramatic to him. The child can be made to relive in its mind the life of the race. Only so can it come to maturity properly equipped for maximum happiness and usefulness. It appears possible that the technique of the education of the future will represent a continuous reaching back from the child's immediate concerns, curiosities, interests, and activities and a continuous coming forward with the integrated, dramatized story of the past. Not until the story had shown the need for man to communicate in writing and the many efforts to produce a method of written communication would the alphabet be taught. Even then, it would be absorbed as part of a great facinating story. It would not be "taught." Motion pictures would show the rude barter of savages, the first crude attempts to transmit information. They would trace the development step by step through pictographs and hieroglyphics to the alphabet used by the race to which the child belongs. The child would see how and why writing developed. He would absorb the story and never need to be taught the arbitrary symbols which, without this understanding, seem so remote from actuality. In the case of man's devices motion pictures would show the evolution of every basic device in such a setting as that from which it must have evolved.

The technique of education of the future will probably be developed with no thought of " teaching." It will, rather, be developed to permit children to absorb unconsciously and without effort, as a seed grows unconsciously and without effort. Discipline has been necessary because the organic conception of knowledge having been disrupted and lost, the interest of the " student " has not been maintained. Seeds do not unfold, absorb, and grow by " discipline " established by agriculture. Minds are like seeds. They will absorb if the material in which they are placed is pleasurable to absorb.[13] The need for discipline is always a proof of the inadequacy of educational technique or personnel for it shows failure to awaken and hold attention. Educational technique needs nothing so much as to emerge forever from the preoccupations both with " disciplines " and with " discipline " which have resulted from loss of vision of the true nature of " knowledge " and " education." The technique of the education of the future will have, at all times, the stirring and satisfying of interests and curiosities as its goal.

In order to be effective in such attempts it will apply continuously the knowledge now coming in ever greater volume from psychology and psychiatry. It will seek ever more consciously and effectively to make the absorbing of knowledge a pleasurable and fascinating pursuit with which no other childish activity can possibly compete. It will certainly be designed to give every child at least as much comprehension of the general outlines and importance of knowledge as globes and maps give world travelers a broad view of the possible places to be seen. It will be certainly free from the present tribal parochialism which so sinisterly infiltrates and colors the primary instruction of all peoples.

IX

IMMEDIATE POSSIBILITIES

The foregoing chapters on the possible aims, material, and technique of the education of the future have dealt with a few basic abstractions seeking to clarify the situation actually confronting the world today.

To the degree in which these abstractions are accepted they might help afford a plan for immediate operation on the part of the few great and powerful agencies charged with the diffusion of knowledge or the new voluntary organizations which may come into being seeking that purpose. It is possible to envisage the success of an " International Association For the Diffusion of Science " to which persons all over the world who have come to believe that this matter is the most important confronting our generation, could contribute; the aims and purposes of which would be to formulate a broad international plan and to seek to have our generation at least initiate efforts looking to its consummation. What is needed is a statesmanship of science. The lack of centralized authority in the old patriarchal or tribal sense means that there is no international authority at present which is *ex officio* charged with the task of integrating and diffusing the knowledge of all the modern world as the tribal leaders of earlier times sought to integrate and diffuse the knowledge of their times. As has been pointed out, the individual scientist is generally too immersed in his own problems to be acutely conscious of his social responsibilities. The international associations of specialists in the various sciences seek largely to bring their respective members together rather than to touch the imagination or life of the general population. Many activities of the great associations for the advancement of science show a developing sense of social responsibility, a developing recognition of the necessity for a statesmanship of science.

It is this task of developing a statesmanship of science which a " World Association For the Diffusion of Science " might well set itself. But already there are great and powerful agencies which might more consciously and deliberately regard their efforts in this light. The present activities of some of these bodies appear to be largely in giving aid to numerous uncoordinated agencies empirically evolved and developed. Whether these agencies can ever, without conscious effort, be integrated into one smoothly-working, national or world mechanism, permitting the continuous diffusion of science which will be of enduring value and use to the individual—touched closely and continuously by various parts of the existing agencies—may well be doubted. Yet conscious effort to integrate all these agencies cannot be vigorous or effective until definite and clearly articulated aims and purposes for which they could be used are formulated and accepted.

Given formulation and acceptation of such aims and purposes, the powerful organizations already charged with the diffusion of knowledge could proceed by plan.

A. They could study the actual and potential need for

and usefulness of every existing agency through which knowledge can be diffused and could rank and grade these agencies so that measures might—over a long period of years—be taken to fit the activities of all the more obviously necessary and useful agencies into one broadly consistent and far-reaching scheme of inter-locking activity.

B. They could seek to make ever more clear to scientists throughout the world the social obligations of science which are not at present recognized. They could keep continuously before the eyes of individual scientists and of associations of scientists the numerous contemporary social aberrations resulting from the failure of the general population to have scientific understanding and, hence, to remain at the mercy of propagandists, bigots, panders, and tricksters.

C. They could direct the attention of the public and of the functionaries in charge of great social agencies—learned societies, universities, colleges; preparatory, secondary, primary, and kindergarten schools; libraries, museums, newspapers, magazines, motion pictures, radio, federated women's clubs, and civic bodies of all sorts—to the rapid progress which could be made in individual, national, and racial well-being if the activities of all these agencies were to be studied as a whole and definite, clear-cut, and wide-visioned plans for cooperation made by them.

D. They could seek continuously to direct the attention of the public to the changed world situation and to the vital need for new educational outlook, new educational philosophy, material, and agencies to meet the new conditions and new requirements of a new age. The public—its attention continuously and effectively directed to such considerations and furnished with new formulations as to the nature of knowledge and the purpose of education—would eventually demand that knowledge be diffused with something of the efficiency that the distribution of material things has reached under modern conditions, and would, almost unquestionably, seek to aid in achieving this end once the possibility of doing so were clear.

E. Efforts could also be made to engender a changed popular attitude toward the idea " knowledge " and to disseminate, through the contemporary technique of publicity and advertising, the thought that the acquisition of knowledge worth the name means individual growth and happiness. The fumbling inefficiency of all the educational devices of today is largely due to the fact that they have to attempt to supply mankind with something which mankind— although giving fulsome lip-service to it—does not especially desire because it seems too formal, remote, and dehumanized. The desire for the appreciation of knowledge can be awakened in humanity precisely as the desire for and appreciation of excellent commodities, devices, or services has been awakened. Knowledge today is obviously cluttered up with all the dreary speculation, mysticism, and metaphysics of the past. When this dead wood is ruthlessly pruned away and the great mass of the public made aware that the new knowledge we have is vital, vivid, and desirable, new ardent desire for it can be awakened with relative ease. With such general desire the existing and evolving agencies for its diffusion can be integrated into an effective mechanism permitting such rapid advancement of civilization as has never before been known.

F. They can serve as—or support—an international agency not merely furnishing educational agencies of whatever nature with financial assistance but furnishing, as well or only, new formulations of their *raison d'être*, their potentialities, and their social responsibilities; their inexorable or desirable relations with the other agencies of civilization for maximum effectiveness or even for the right to exist.

No activities of this nature are at present being deliberately undertaken. They cannot be undertaken until science and education evolve their own statecraft. The logical agencies for the beginning of this evolution are the great educational endowments which have, for the first time in history, an opportunity to deal with the problem of the diffusion of knowledge from a completely international viewpoint. If no more were done by these existing foundations than to aid in the organization and development of a "World Association For the Diffusion of Science" the first realistic efforts at the great undertaking now possible to mankind would be made. For there can be no doubt that if such a world association were to be brought into being and were to have on its directorate the outstanding scholars and scientists of the modern world it could quickly coordinate all the numerous agencies which are at present relatively inefficient for lack of coordination. It could serve as an international body to marshall the aspiration, idealism, and social purpose of myriads of men and women who cannot give their minds to any of the present creeds, doctrines, parties, and economic or political movements.

Such a "World Association For the Diffusion of Science" could act as the agency for bringing together the international commission which could plan a minimum curriculum and, because of its prestige, could bring this minimum curriculum to the attention of every centralized social agency on earth.

Through its developing activities, far-reaching plans and programs designed to be carried on and perfected over generations or centuries could be initiated—plans and programs designed for the production and elaboration of such efficient machinery for the diffusion of knowledge that all the agencies of human intercommunication could be—together—regarded, by a happier future, as parts of a world university affording instruction to every individual from birth to death; bringing and keeping every individual within the circulating blood stream of human thought and making mankind the conscious organism which it must some day become.

NOTES

Note 1, Page 31. The following quotations appear to give abundant academic support to these statements:

1. "Over-specialization is the greatest defect in our colleges, and I believe also in our entire field of public and high school education. I consider that a good engineer should not know steam engines to the exclusion of all else, but that he should have a background of history and literature and political economy, so that he can link his engineering world up with the rest of mankind. If he knows nothing but steam engines, he forgets all else about him. He forgets that the success of democracy is dependent upon his participation in it; that he has an obligation to his neighbors. We are just beginning to reap the effect of this over-specialization. I refer now to the wave of prejudice and perverted thinking which has swept this country. At no time since the Civil War have serious men been more worried over the soundness of American Democracy.

    " The college student is the most conservative person in this country. At least, I consider him so. I sometimes wish that he were less reactionary. Specialized and technical educations do not broaden our citizens into great leaders."—*Dr. Livingston Farrand* (President of Cornell University).

2. "Specialization, usually considered the foe of culture and the friend of science, has dangers also for science.

    " I can imagine generation after generation turning out scientific workers with narrower and narrower outlook, incapable of conceiving those broad flights of imagination which have made possible every epochal advance in science.

    " The classics have been killed by classroom pedants, who have forgotten literature in being absorbed in the minutiae of their specialty.

    " How can we protect specialists against fragmentation of background? It is a national issue of the first magnitude to protect the safety and sanity of the social order. Universities must be a real training ground for political and industrial statesmen who have perspective as well as power."—*Glenn Frank* (President of the University of Wisconsin).

3. " Efficiency is a fine by-product of education, but to make efficiency

the object of education is to debase that fine thing which we call character.

" For many years we have been greatly influenced by German educational methods, not realizing that the educational process in Prussia at least is designed to promote efficiency. Is this the difference between kultur and culture? It is a serious tendency which we observe in college catalogues of the present time—this tendency to use the precious four years of college to enable a man to get a living. Those years should be devoted to making living worth while.

" But you will ask, How is education, the process of education, this life-long process of education, to be assimilated to character? Let biology answer us—by functioning. The generous use of knowledge and training in promoting the well-being of mankind will return to us in character, in ever-growing high manhood, in satisfactions that perish not, in those qualities of being which live on forever, because they are life."—The late *Dr. Wallace Buttrick* (of the General Education Board).

4.   " Meantime, at the college level, many institutions, Yale among them, are making every effort to offset some of the defects of the earlier part of the present system of stimulating intellectual individuality and independence and by recognizing maturity of mind and according to it *bona fide* freedom of opportunity. To be sure, at the college level it is too late to save the student the years he has lost, but at least we can make the attempt to furnish a less flabby and superficial training than has often been the result achieved by our education. We can do it partly by stiffening the actual intellectual requirements, and partly by stimulating the student to a more fundamental use of his intelligence; and it is to this end that our tutorial systems, our comprehensive examinations, our honor courses and the like are directed."—*James Rowland Angell* (President of Yale).

5.   " We must stop trying to teach, trying to instruct, because it can't be done. We must take the responsibility for learning from the teacher and place it with the person who is commissioned to learn. It is silly business trying to give to students a set of conclusions. No one ever studied who knew the answer in advance."— *Dr. Alexander Meiklejohn.*

Note 2, Page 39. Einstein, in his message to Edison (on the latter's eightieth birthday) said:

" The technical geniuses of the world, of whom you are one of the most successful, have confronted mankind with a new situation in the course of the last half century—a situation to which it has not yet succeeded in adapting itself."

Note 3, Page 50. The opinion of contemporaries is, fortunately available. Charles Francis Adams, president of one of the great trans-continental railroads, grandson of a President of the United States, son of an ambassador to England and father of a Secretary of the Navy, wrote in his autobiography:

" I have known, and known tolerably well, a good many " successful " men, " big " financially, men famous during the last half century, and a less interesting crowd I do not care to encounter. Not one that I have ever known would I care to meet again either in this world or the next. Nor is there one of them associated in my mind with the idea of humor, thought or refinement."

In the *Commonweal* of January 16, 1928, Mr. Charles Willis Thompson writes that Theodore Roosevelt, while President, once said to him:

" I have to talk to millionaires but I wish I didn't. They bore me. You'd suppose the master of a great industry would be full of interesting things to say. But they're not. They know their own businesses, but the moment they stop talking shop they haven't an idea. Outside of money making they're dumb. I don't mean that as a unanimous condemnation but it applies to nearly all of them."

Note 4, Page 57.

" I look forward to the day when we shall have a system of adult education in the state which will reach every man and woman as we are now reaching the child. And the librarian will be as important a factor in that place as the formal teacher or the lecturer, perhaps the most important and inspiring factor."—*John H. Finley.*

Note 5, Page 61.

" If you could get but one result out of your university course, I would rather you would get a love for and a habit of reading good books than any other. With such a feeling and habit you will become educated and cultured men and women; and without them there is little if any chance of your doing things."—*Herbert S. Hadley.*

Note 6, Page 86.

" Education cannot be left to bankers, politicians and tradesmen or any other type of citizen whose prime interest is not education."—

Dr. J. E. Kirkpatrick ("The American College and Its Rulers." New Republic, Inc.).

Note 7, Page 93.

"That a free community book exchange is destined to be transformed into an active intelligence center through the addition of a competent staff of scholars trained in fitting books to human needs, is an idea as dimly perceived today as was the free library itself seventy-five years ago."—*Dr. William S. Learned.*

Note 8, Page 97. The reader is very especially referred to two papers by Count Korsybski in which his ideas are developed:

Time-Binding—The General Theory (E. P. Dutton)
Time-Binding—The General Theory (Second Paper) (James C. Wood, 2021 Nicholas Avenue, S. E., Washington, D. C.).

While the basic ideas in this present book have been expressed by him long before these papers were known, the author wishes, nevertheless, to make explicit acknowledgement that Count Korzybski's papers have greatly influenced some of the definitions given and some of the ideas expressed.

Note 9, Page 120. In this connection the evidence given by Thomas A. Edison in *re* Federal Trade Commission *vs.* Famous Players, Lasky Corporation *et al.*, May 15, 1923, is especially interesting:

Q. Have you made any investigation which has resulted in your making estimates as to the relative value of the different senses as evidenced in obtaining information, if so, what are those estimates?

A. Yes, I have made a considerable number of investigations in that line.

Q. Will you explain what they are, what those investigators were?

A. Well, I got in my head that it would be a fine thing to teach children by that method, teach them by pictures, because they resent books, they do not like them, and they will look all day at moving pictures, so I started in a number of years ago, put in a department and I had a visionary scheme of putting it all over among the schools of the United States, and I made a lot of pictures to teach chemistry and physics, and things like that, those complex things, to children from six, eight and ten years of age. Well, I made some to see how they worked. I got twelve children, boys and girls, and I tried it, as they say, on the dog. I had their mothers when the children were not home, and I showed them two, three little pictures and I had their mothers write down what the little ones remembered and understood and I was quite surprised that they understood a great deal, and those parts which they did understand, I made them over again and tried them perfectly. Then I made some other pictures and I think I got as high an understanding of those little people of about 80 per cent the first time. Then I saw that you could teach them anything. Of course, I made it very attractive, I put a little boy and a girl in my technical experiment, I had a little boy and a girl in the kitchen doing all the experiments. Of course, I trained these people and that gave them confidence, they could understand it, and they certainly did, and it is very remarkable. I think they are better than men after, they are more susceptible, and I thought, of course, I would like to put it into the schools and I was going to make pictures for them and I had the Board of Education over there, I think about ten people, and they were all delighted and came back to New York and never did anything. Then I tried it on other school people, but I found I was soon up against the book publishers that publish books for schools, and I saw that I would be defeated so I quit. Well, I think we learn 80 to 90 per cent through the eye and through books, and it is so easy to do it with the eye. You can teach without the person suspecting he is being taught, that is the great thing. If these children thought they were being taught I don't know whether they could understand any of it, but they did not know it. We weave it into something interesting and they are taught all those things and do not know it. We are not taught much through the ear except music, all the eye.

Q. What effect does this have upon the potential power of the motion picture in disseminating knowledge.

A. Well, in my opinion the moving picture is 100 per cent perfect for teaching anything. For instance, they have a bottle machine for making beer bottles and all kinds of bottles and it about fills up all this room, very complex, and the inventor did not have very much money, but he had one machine, and he could not get people to come and see it from Europe and other places. He came over to me and I made him a moving picture showing the exact process of making bottles, in fact, him making the bottles. He took that moving picture over to Europe and sold $80,000 worth and he got the cash on those things. Now, things like that. It is very valuable. But for schools, that is the great place, and for the general public. You can bring out anything. It can be made to improve the morals of the people if you want to, or you can do the opposite.

Q. What effect does it have upon the conduct and the taste and manners of the people.

A. Well, if you know if Darwin was alive—we are very imitative and the young people and some of the elders are liable to imitate anything they see of that kind in a moving picture. They have not very much imagination themselves, so to supply their imagination through the pictures, they absorb it and it is a part of their lives and they carry it out.

Q. What, in your opinion, will be the future growth of the motion picture?

A. I think it is just about started. Of course, we are out for millions now, but I notice in the trade papers I take and the educational papers that it is spreading all over. I am going to attend a dinner tonight of the Moving Picture Chamber of Commerce, I think, something of that kind, at Delmonico's, non-theatrical, and there are a lot of concerns going into that business and I think in 20 years you will find that all the children will want to go to school; they will not get behind, there will not be any truants, if you make the pictures they like, and we will have plenty of high-brows.

Note 10, Page 121.

" Apart from the modicum of technical instruction they impart, the upper schools and universities of our world already betray themselves for an imposture, rather delaying, wasting and misleading good intentions, rather using their great prestige and influence in sustaining prejudice in favor of outworn institutions and traditions that endanger and dwarf human life, than in any real sense educating.

They are the most powerful bulwarks, necessarily and inseparably a part—the most vital and combative part—of that declining order which our revolution seeks to replace from the foundations upward. Here, as with monarchy and militant nationalism, we do not need so much to attack as to disregard and neglect, to supersede and efface, through the steadfast development of a new world-wide organism of education and interchange, press, books, encyclopedias, organized translations, conferences, research institutions.

A time must come when Oxford and Cambridge, Yale and Harvard will signify no more in the current intellectual life of the world than the monastery of Mount Athos or the lamaseries of Tibet do now." H. G. Wells (in The Cosmopolitan, September, 1926).

Note 11, Page 121.

" There is nothing occult or supernatural in the processes of life, and eventually we will unravel its secret. Protoplasm is nothing but a chemical compound. I see no reason why some day we shall not be able to produce it. When we do so it may be living or it may be dead, no one can say."—*Dr. Paul Renno Heyl.*

Note 12, Page 124.

" The costly, ineffective and ever demoralizing character of much contemporary school and college work is due to the fact that so many of those who conduct it can neither look back down the road over which mankind has come nor forward along the road over which mankind is moving.

" They live in a state of unstable equilibrium, without cognizance or appreciaton of those ideas which silently and unconsciously shape and guide the action or the inaction of men.

" The exclusive and intensive study of natural science, now carried on more than a full generation, has made no impression whatever upon the public mind. That mind continues to come to its conclusions and to formulate its choices with serene unconcern as to whether any such thing as scientific method exists. The ability to read has well nigh disappeared if the reading be serious, instructive or ennobling; the ability to write, so far as it exists at all, delights to manifest itself in forms of exceptional crudeness and vulgarity; the ability to perform the simplest mathematical operations is, to all intents and purposes, confined to teachers of mathematics or to specialists in that subject."—*Nicholas Murray Butler* (President of Columbia University).

Note 13, Page 126.

" Human nature isn't so stupid after all; and if a healthy boy or girl finds school work a bore, it means, in most cases, simply that the particular type of artificial experience which we call class instruction is not likely to be profitable for that particular person. A suitable test might bring out real capacity for learning certain manual skills in the case of a person for whom sound book learning would be a waste of time."—*Dr. Frederick Keppel.*